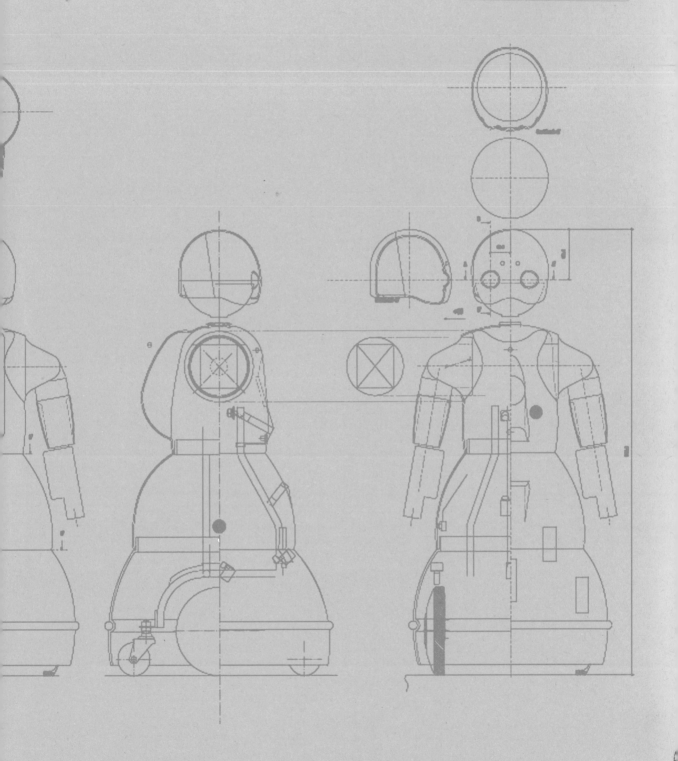

LOVING THE MACHINE

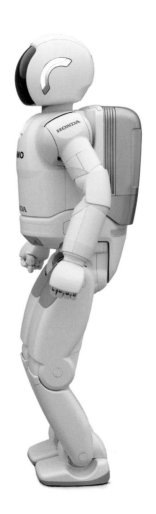

LOVING THE MACHINE

The Art and Science of Japanese Robots

Timothy N. Hornyak

KODANSHA INTERNATIONAL
Tokyo • New York • London

Distributed in the United States by Kodansha America Inc., and in the United Kingdom and continental Europe by Kodansha Europe Ltd.

Published by Kodansha International, Ltd., 17–14 Otowa 1-chome, Bunkyo-ku, Tokyo
112–8652, and Kodansha America, Inc.

First Edition, 2006
15 14 13 12 11 10 09 08 07 06   10 9 8 7 6 5 4 3 2 1
        Library of Congress Cataloging-in-Publication Data

Hornyak, Timothy N.
    Loving the machine : the art and science of Japanese robots / by Timothy N.
Hornyak.-- 1st ed.
        p. cm.
    ISBN-13: 978-4-7700-3012-2
    ISBN-10: 4-7700-3012-6
    1. Robotics--Japan--Design and construction. 2. Androids--Japan--Design and
construction. I. Title.
TJ211.H66 2006
629.8'920952--dc22
                            2006002851

www.kodansha-intl.com

CONTENTS

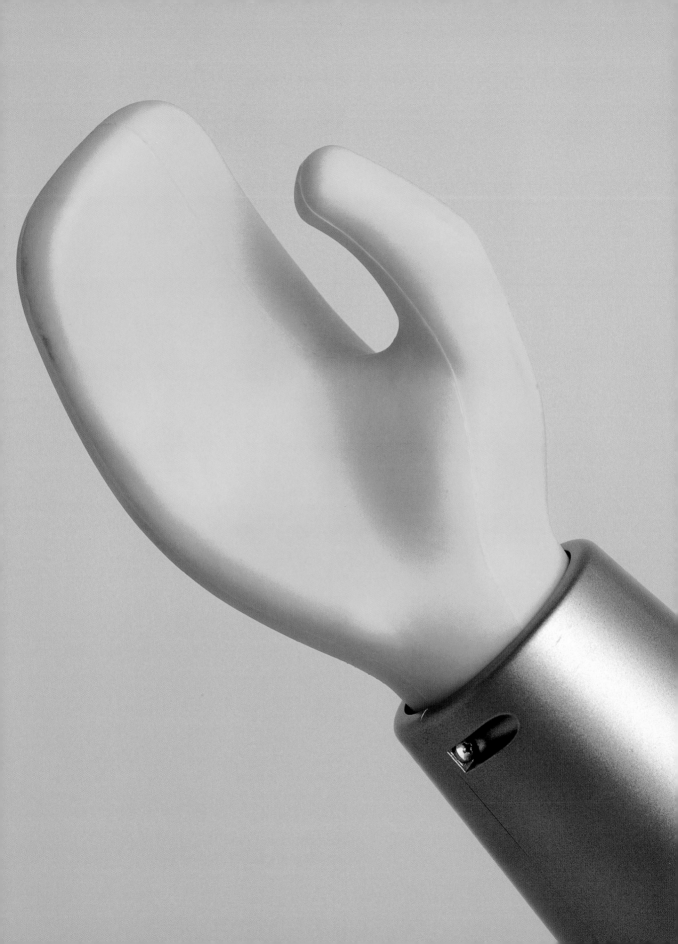

Say Hello to the Future

Welcome home.
　　—Wakamaru

I met the robot in the heart of Tokyo, a stone's throw from the Imperial Palace. It was a crisp autumn evening, and bell crickets trilled in the darkness over the Hibiya moat as I walked to the furniture store. On the second floor of the nondescript building amid some designer beds and tableware was Wakamaru, a household robot named after a Japanese warrior who lived eight centuries ago. It had insectile but expressive eyes and a bright lemon yellow body.

I quickly found that talking with a robot can be disconcerting. In Japanese, I asked it about the weather. Tokyo skies would be cloudy with a chance of rain, I was informed. (Wakamaru knew this due to its wireless Internet hookup.) It said this in a voice neither male nor female, mechanical but tinged with the solicitude one hears in automated public announcements in Japan. Its Japanese was formal though not excessively ceremonious. And then the salespeople took over, demonstrating how the robot can greet its owner returning home or remind one to take an umbrella when going out in inclement weather.

Wakamaru can also read one's horoscope for the day, they explained, which it quickly proved by launching into an account of things I should take note of: "Regarding relationships," it said, "you tend to use bitter words to the one you love. It's better to exercise moderation, using sweet, tender words to that person." I asked it the time, which it dutifully announced, but I was quickly running out of conversation topics. The robot stood waiting, silver arms hanging by its hoop-skirted lower half. It regarded me

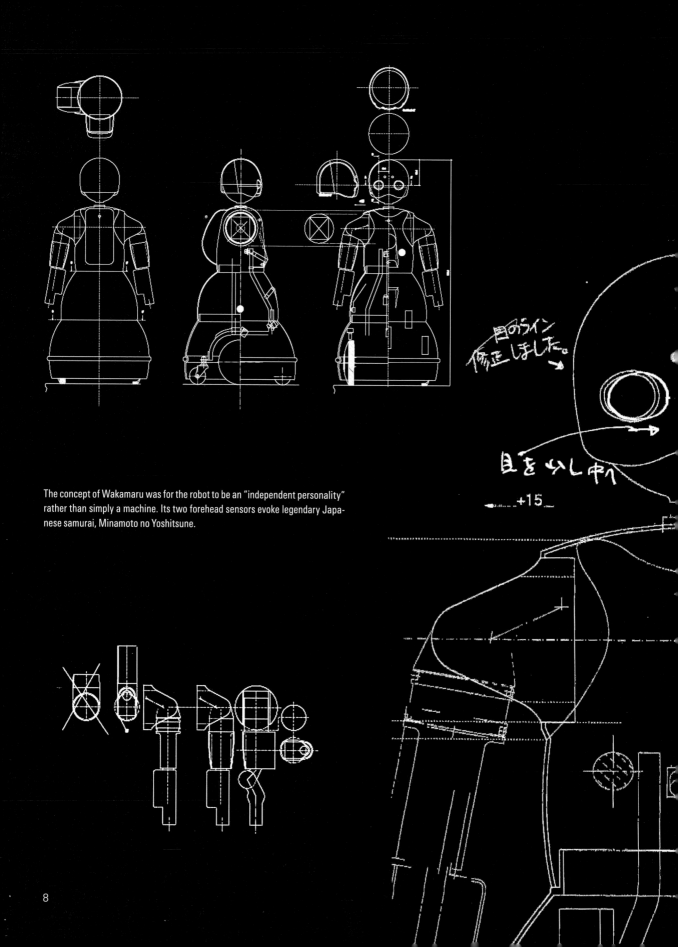

The concept of Wakamaru was for the robot to be an "independent personality" rather than simply a machine. Its two forehead sensors evoke legendary Japanese samurai, Minamoto no Yoshitsune.

目のライン
修正しました。

目を少し中へ

+15

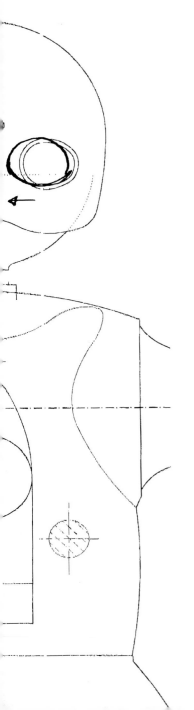

silently with its big black eyes—which were beginning to look more puppy dog-like—and its gently arching mouth gave it a slightly pitiful air.

The bubble head angled upward, shifting to follow me as I shifted position. Ah, it's face-tracking, I thought. I knew from my research that two cameras in the robot's head were monitoring its surroundings and the images were being processed by algorithms allowing it to detect faces and follow them. I knew that this thing looking up at me was, after all, just a machine: a copolymer plastic head, two arms, a wheeled base, sensors, motors, a CPU, control circuits and a battery.

Yet, while staring back into its eyes, something strange happened. I began to feel its awareness of me on an almost cellular level—and the hairs tingled on the back of my neck. The gaze was penetrating, the eye contact unmistakably animal. No, it was human—like a child looking up at a parent. Looking up at one of his own. The urge to accept this humanoid robot in front of me as something more than just complex clockwork was irresistible. Something that analyzes, absorbs, learns. Something that reacts, something that thinks.

Something that *is*.

Wakamaru, a Mitsubushi Heavy Industries household helper for sale in Japan, is one unit in a robot army marching into the country from university labs, corporate research centers and toymakers. It joins legions of automatons both real and imagined that have populated factories, toy store shelves and the pages of comic books for decades. These newly developed robots, designed to serve ordinary consumers, appear almost daily in Japan. And people, it seems, can't get enough of them.

At a supermarket in Chiba Prefecture, Fujitsu's latest humanoid robot starts a new job as a store clerk. Enon, four feet three inches tall, offers shoppers guidance and advice. "It's chilly today," it tells one young customer, "How about some spicy food to warm up?"

In mountainous Gunma Prefecture, meanwhile, an engineer at an environmental machinery maker brings the dreams of millions of Japanese boys of all ages a big step closer to reality by building the Land Walker, a one ton, eleven-foot-tall steel cockpit and pair of legs that evokes the colossal spacefaring warrior

Wakamaru's red LED chest display can show a variety of icons, such as an envelope indicating email. An example of its practicality is its wireless Internet connection—it can read out messages or email its owner if it detects anything unusual while house sitting.

piloted by humans in the popular *Mobile Suit Gundam* animated series. It too fires projectiles—spongy pink balls. The Land Walker's top speed: one mile per hour.

In Nagoya, a local venture company unveils a robot version of the perennial icon of the Japanese fetish for all things cute. This robot cat can remember faces and has a pink ribbon that lights up during voice recognition. Even though Hello Kitty Robo lacks a mouth like the original cartoon feline, it can engage in simple conversation and remember names.

In a nursing home on the other side of the country, a caretaker finds that an elderly resident has snuck something soft, cuddly and white into bed: a robot seal. The resident's bed partner is Paro, a therapeutic artificial pet that can respond to voices and stroking. Not only do patients accept the baby harp seal robot, even forming emotional attachments to it, the fuzzy machine reduces their stress levels and depression.

In the capital, a robot with dazzling dexterity and panache stands up on a floodlit dais, its yellow eyes ablaze, and leads the largest orchestra in Japan through Beethoven's Fifth Symphony. Under the baton of Qrio, a walking, talking, silver gnome-sized humanoid created by electronics titan Sony, the Tokyo Philharmonic Orchestra thunders through the famous opening bars, dozens of human musicians in sync with their magnesium-bodied conductor, which waves its arms with marvelous fluidity. It is the first time in the world for a humanoid robot to conduct an orchestra in a live performance. And little Qrio pulls it off in style.

Japanese robot developers are nothing if not big dreamers. Their unswerving ambition aims at goals such as creating a team of robot soccer players that will be able to beat the reigning World Cup champions by 2050, and even creating a full-fledged robot boy with the physical and mental faculties of a five-year-old child. The government sees humanoid robots as a priority R&D field that will safeguard the country's status as a scientific and technological powerhouse, and provide caregivers for the elderly in a rapidly aging society.

The powerful Ministry of Economy, Trade and Industry envisions Japan becoming a "Neo Mechatronics Society" within a decade, with robots fully integrated into everything from taking out the trash to forestry management. It believes that by 2025 the domestic robot market will swell to ¥6.2 trillion, a figure

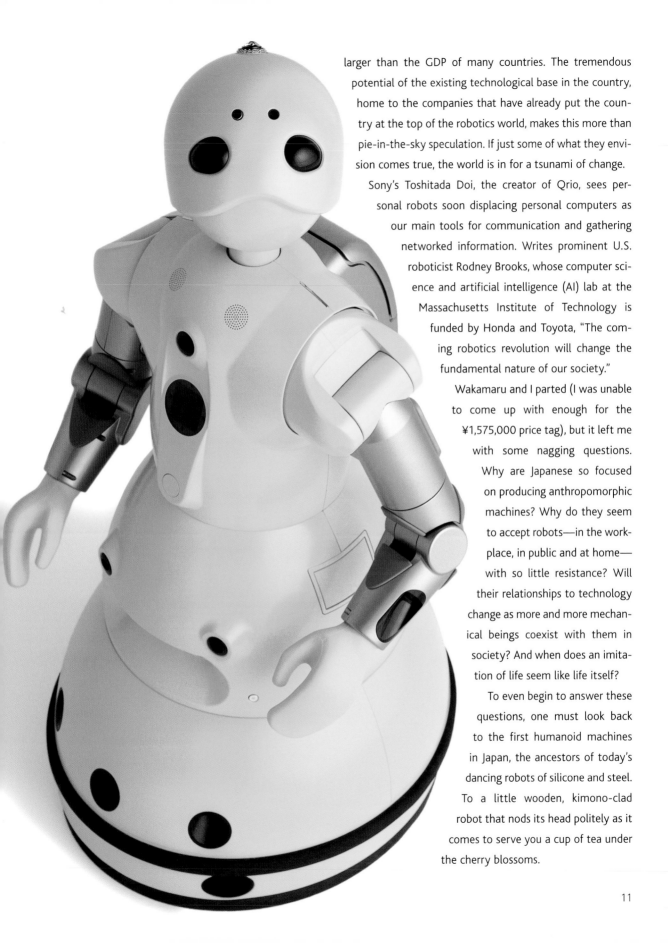

larger than the GDP of many countries. The tremendous potential of the existing technological base in the country, home to the companies that have already put the country at the top of the robotics world, makes this more than pie-in-the-sky speculation. If just some of what they envision comes true, the world is in for a tsunami of change.

Sony's Toshitada Doi, the creator of Qrio, sees personal robots soon displacing personal computers as our main tools for communication and gathering networked information. Writes prominent U.S. roboticist Rodney Brooks, whose computer science and artificial intelligence (AI) lab at the Massachusetts Institute of Technology is funded by Honda and Toyota, "The coming robotics revolution will change the fundamental nature of our society."

Wakamaru and I parted (I was unable to come up with enough for the ¥1,575,000 price tag), but it left me with some nagging questions. Why are Japanese so focused on producing anthropomorphic machines? Why do they seem to accept robots—in the workplace, in public and at home— with so little resistance? Will their relationships to technology change as more and more mechanical beings coexist with them in society? And when does an imitation of life seem like life itself?

To even begin to answer these questions, one must look back to the first humanoid machines in Japan, the ancestors of today's dancing robots of silicone and steel. To a little wooden, kimono-clad robot that nods its head politely as it comes to serve you a cup of tea under the cherry blossoms.

CHAPTER 1

Clockwork Teatime

*The wheels
of a doll serving tea—
working hard*
　　　—Ihara Saikaku

The beach on Kinuura Bay is sweltering under an unseasonably fierce May sun. Early arrivals are zealously defending the shade provided by a few slim trees at the edge of the sand, and police intent on crowd control shuffle the steadily increasing horde along. By mid-afternoon they have filled the jetty and a cordoned-off zone along this narrow inlet in Japan's industrial heartland south of Nagoya.

Around two o'clock, a loudspeaker punctuates the onshore breeze to announce the start of the ceremony. The spectators stand, children on shoulders and cameras poised, craning their necks toward the road between the beach and Kamisaki Shrine. There is a sound of flutes, drums, samisens and an undulating sea song that heralds the appearance of hundreds of men. They are clad in white headbands and tabi socks and colorful robes with diamond, wheel and gourd patterns, and fat Chinese characters emblazoned on the back. They shout as they strain at their places along long thick tow ropes.

They are pulling a *dashi*, or festival float: four tons in weight and twenty feet tall, with two wooden decks, crimson drapes embroidered with flaming orbs and tall poles with banners bobbing in the wind. It towers over the tanned young men leaning into the wooden bars just over the wheels as they drive it onto the sand, where it grinds to a halt.

While the flutes continue their high plaintive strain, the crowd

Dating back centuries, the annual Kamezaki Shiohi Festival near Nagoya is a colorful pageant of craftsmanship and traditional puppetry. After being paraded through the community and ritually cleansed in the sea, five enormous *dashi* wooden floats are lined up before Kamisaki Shrine (opposite). The floats feature *karakuri* puppets that portray stories and dances from Japanese history and folktales on ornate stages (this page). The Tanaka float crew's Puppeteer (above) even puts on a play within a play with miniature puppets that perform the Noh theater dance drama *Funa Benkei*.

As these prints from the Edo Period (1600–1867) attest, clockwork automata were a source of delight and wonder for both Japanese and foreigners. The *Settsu Meisho-zue*, a sightseeing guidebook to Settsu Province (old Osaka), shows a group of Dutchmen (right) who are absorbed in a Takeda stage performance of karakuri actors in *Funa Benkei*; one of the foreigners is so shocked that he has forgotten to sit down. Another print (above right) shows a group of nobles enjoying *zashiki karakuri*, or room automata, including a tea-serving doll.

takes a closer look. The carriage is constructed nearly entirely of fine woods like zelkova and ebony, as well as lacquer, brass and gold, with graceful bowed roofs and elaborately carved railings depicting turtles, birds, dragons. About ten boys also clad in traditional festival attire are riding shotgun on the top deck. Their chariot might be mistaken for part of a Buddhist temple that has been borrowed for a joyride, but its main feature is something of a more secular nature: a theater at the front, complete with a curtain, stage and patiently waiting actor.

Four other floats follow. They date from the nineteenth century and have names like Blue Dragon and Flower King. Each in turn is laboriously hauled through the low tide for a ritual cleansing to the cheers of the crowd. After being pulled up to the shrine, and ceremoniously paraded in circles amid much chanting and jostling, their exhausted, sweating handlers adeptly line them up a few inches apart, facing the sanctuary. With the fad-

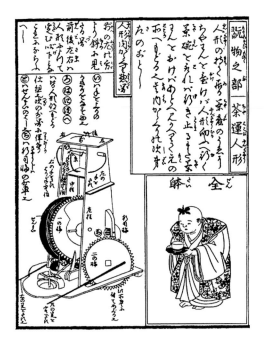

The three-volume 1796 treatise *Karakuri Zui* ("Illustrated Machinery") is an extraordinary blueprint detailing how to make karakuri automata and other devices. Former Waseda University professor Shoji Tatsukawa, who used it to recreate a tea-serving doll in the 1960s after the know-how had been lost, says, "The most difficult challenge was producing the doll's main cogwheel. The key was using several different pieces of wood so it would not crack."

ing light glinting off their gold and brass trimmings, the dashi look like a row of baroque cuckoo clocks waiting to chime.

Suddenly and simultaneously, plays begin on these mobile stages. To the sound of wailing flutes and slow, punctuating drumbeats, scenes are enacted from myths and legends. One stage shows the fairy tale of Urashima Taro, a fisherman who rescues a turtle and is rewarded with a visit to the Dragon King's undersea palace. On a neighboring stage an acrobat in old-fashioned Chinese clothing pirouettes through trapezes under a canopy of cherry blossoms. Warriors of fierce visage twirl spears, and a pair of clownish figures crank a clockwork mechanism while another does handstands above.

Elaborate backdrops change along with the actors' masks as the stories progress. The exquisite costumes rival those of any Noh or Kabuki stage and the craftsmanship of everything from fans to headdresses is equally impressive. The movements of the actors are complex and symbolic, rough and graceful. One is tempted to say, almost human.

In fact, all of the actors performing in this centuries-old show are robots. Or—more accurately—very early Japanese robots. They are kimono-clad wooden puppets manipulated by invisible yet fiendishly complex means.

This is the Kamezaki Shiohi Festival, a ritual dating back to the late fifteenth century. Aside from the ostentation of the carriages, the festivities are renowned for their diminutive stars, who enchant with their gorgeous costumes, acrobatics and stagecraft. These amazing puppets are known as *karakuri*, meaning "trick," "mechanism," or "gadget." The trick is how the karakuri seem to act independently. The gears, rods, silk cords and puppeteers controlling them are concealed in and under their bodies, and watching the little acrobat twirl through the bars without visible human intervention is pure magic. These automata seem autonomous.

As the stage shows draw to a close, boys from the five floats gather in front of one of the dashi. On the upper stage, a karakuri version of the Noh play *Funa Benkei* is being acted out. It's the tale of how the tragic hero Minamoto no Yoshitsune—after whom Mitsubishi's household robot Wakamaru was named—is confronted by the vengeful spirit of a fallen enemy. Remarkably, the show is actually a play within a play, for the action takes place on a wooden box that has partly opened on the knees of a

large, jolly-faced puppet called The Puppeteer. At the conclusion of the drama, The Puppeteer fires a spring-powered toy weasel into the cheering crowd, signaling an end to the revelry. Locals believe that the boy who catches it will be blessed with good luck for the following year.

The Puppeteer and its karakuri are the sole remaining act from a line of clockwork masters who were largely responsible for popularizing these karakuri dolls as a form of mass entertainment in Japan. Indeed, their mechanical heritage is part of the forces that helped pave the way for Japanese industrialization in the nineteenth century and the birth of true robots in the next.

Impresario and Missionary

In 1662, a businessman and impresario named Takeda Omi opened an amusement park by a canal in Osaka's thronging Dotonbori entertainment district. His Takeda-za theater staged experimental performances featuring astounding automats as well as human actors and puppets. The novelty of the outdoor cabaret, which included devices powered by water from the canal, resulted in a smash hit. "The shows had cheap fares and generated lots of attention," says Kazuhito Yamada, a literature professor with a deep interest in traditional puppet plays. "People from across Japan came to see the karakuri and were stunned."

Takeda's mechanical men were such a success that he was nicknamed Takeda Karakuri. His secular puppet dramas became a genre unto themselves and influenced other forms of Japanese performing art. A saying of the time even held that "You haven't seen Osaka if you haven't been to the Takeda karakuri."

Even the businesslike Dutch, who had special permission to trade at Nagasaki port during the Edo Period, weren't immune to the charm of karakuri. The illustrated *Settsu Meisho-zue*, a sightseeing guidebook, shows a group of Dutchmen absorbed in a Takeda stage performance of karakuri actors in *Funa Benkei*; one of the foreigners is so shocked that he has forgotten to sit down.

With his fame, Takeda Omi founded a lineage of doll and puppet master craftsmen, often apprentices who took his name, that continued into the late eighteenth century. They opened theaters in Edo (old Tokyo), Nagoya and other cities, building wondrous dolls that employed clockwork mechanisms, weights, stoppers, gears and springs. The eighteenth-century *Dai karakuri-ezukushi*

◀ When the *chahakobi ningyo*, or tea-serving doll, was recreated in the 1960s, it was hailed as a "robot from the Edo Period." Dressed in an elegant jacket and *hakama* pleated pants, this doll was made by karakuri master craftsman Shobe Tamaya IX and is part of the collection of Toyota National College of Technology Dean Yoshikazu Suematsu.

At his atelier in Inuyama, north of Nagoya, karakuri craftsman Shobe Tamaya IX demonstrates his skills and creations to visitors. Tamaya is part of a lineage of karakuri artisans dating back to the eighteenth century. "It's pretty difficult," Tamaya says of his craft. "As a karakuri doll maker, you have to make all the parts yourself, from the beautiful stands to the clothes they wear. As a pupil undergoing training, it usually takes fifteen or sixteen years before you can produce a complete doll."

("The Great Picture Book of Automata") features illustrations of Takeda karakuri that write Chinese characters with their hands or mouths or do handstands on trapezes. In one Takeda show around 1750, dolls shot blow darts at fans, climbed ladders and even urinated on stage.

Takeda Karakuri's amazing automatons, however, would most likely never have existed without the entrance of Western gun and clock-making technology to Japan in the middle of the sixteenth century. The Spanish Jesuit missionary Francis Xavier is believed to have introduced the first Western clock when he presented it to a feudal lord in 1551. Later Christian missionaries taught Japanese how to make clocks, organs and astronomical tools at a vocational school called the Seminario that was established in Nagasaki around 1600. Other Japanese craftsmen took imported timepieces apart, figured out how they worked and copied them; the country's first mechanical clock was created in 1605.

This was during the first years of the Edo Period, at a time when Japan was finally at peace after centuries of internecine wars among rival samurai clans and warlords. In order to maintain its grip on power, however, the Tokugawa shogun regime began implementing a series of policies designed to prevent any challenge to its authority. One was the policy of *sakoku*, or national seclusion, which in 1639 banned overseas travel and trade; only Dutch and Chinese merchants were allowed at Nagasaki. Japan would be largely cut off from the world for the next two hundred and fifty years.

With no input from abroad, Japan's clockmakers were left to their own devices. Japanese clocks, known as *wadokei*, became large, ornamental affairs to grace the mansion of a samurai lord, and they required intricate adjustments. They told the time by the traditional Japanese method—in which night and day were each divided into six segments that were named after animals. Further complicating matters, the segments were of unequal length and varied seasonally, so ingenious mechanisms employing weights and sand were used to compensate. Some wadokei measured time by the burning of incense sticks, while others were far more sophisticated. One type called *Yagura-dokei*, or turret clock, was mounted on a high stand to accommodate long hanging weights, and featured advanced functions such as chimes and calendars along with a stationary hand and rotating dial.

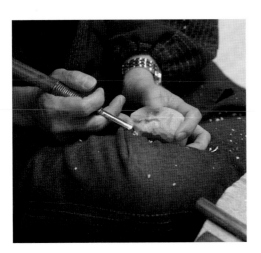

"The most difficult is the face," Tamaya says. "People's reactions will really change depending on whether the face is good or bad. And each doll's face is unique."

Ultimately, wadokei were more beautiful than accurate; the best are works of art rather than useful machines. Their complexity and cost meant they were known mostly to the rich but the technology they incorporated allowed the development of Japan's most sophisticated and striking karakuri automata. The best illustration of this is an amazing treatise from the eighteenth century that contains blueprints for one of the world's oldest mechanical robots—a karakuri doll that serves tea.

Enter the Social Machine

Quick and precise, Shobe Tamaya's hands shape a pale globe of Japanese cypress. The wood turns in his palm, a flurry of shavings drift to the tatami mats on which he kneels, and—as though approaching through a mist—a face begins to appear. He has done this countless times, but all his attention is locked on the object in his lap. Another karakuri doll is coming to life.

Tamaya is the ninth generation in a line of karakuri craftsmen and artisans that stretches back to 1733, when Shobe Tamaya, a Kyoto puppet master, moved to Nagoya. He and his descendants repaired and rebuilt the mechanical puppets used on floats in the many religious festivals of Aichi and Gifu prefectures. In fact, a Tamaya karakuri from the 1780s is still in perfect working order in Nagoya. Today, 52-year-old Shobe Tamaya IX is one of Japan's few remaining traditional karakuri craftsmen.

The face he is working on is by far the most famous karakuri face: a chubby boy with the pallid features of a Noh mask. Proudly displayed on Tamaya's workbench is a completed version of the figure. It is a *chahakobi ningyo*, an eight-inch-tall, windup doll clad in kimono and *hakama* pleated skirt that can perform the quintessential act of Japanese hospitality—serving green tea. When the host places a cup on the tray held by the doll and faces it toward a guest, the karakuri travels the appropriate distance preset by the host on its two hidden cogwheels, courteously offering the tea, and nodding its little shaven head all the while. It halts when the cup is taken from the tray. After the guest drains the cup and replaces it, the doll turns straight around and returns to the host.

This unique interactivity makes the windup tea-serving doll a social machine, in which the main purpose, like the Japanese humanoid robots that are its natural descendants, is communication with human beings.

First appearing in the mid-Edo Period, the little doll was widely known and loved by the public. The haiku poet Kobayashi Issa (1763–1827), even dedicated verse to it:

Such coolness by the gate
as the tea-serving doll
brings another cup

Modern writers have also sung its praises. "The doll is modeled in the form of an adorable child rather than the machine-like form of a Western robot," writes renowned architect Kisho Kurokawa. "In the Edo Period, the technology that was incorporated into a device was not displayed on its exterior but incorporated invisibly in the interior, giving people a feeling of wonder and mystery. The role of machines was not to express their own independent identities but that of human beings."

Recreating the doll would be impossible without the existence of the three-volume 1796 treatise *Karakuri Zui* ("Illustrated Machinery") by Yorinao (Hanzo) Hosokawa, a polymath and inventor who was also an artisan and mechanical engineer. Born in the feudal domain of Tosa on Shikoku Island, Hosokawa learned architecture and carpentry from his father. He studied astronomy and Confucianism in Kyoto, where he was credited with creating an astronomical globe and pedometer. He was even involved in reforming Japan's calendar under the shogunate.

When he left rural Shikoku for the sophistication of Kyoto, he is said to have inscribed a vow on a hometown bridge promising not to return until he had become famous; his nickname of Karakuri Hanzo suggests he did just that. His brilliance is abundantly clear in the woodblock-printed treatise that meticulously details how to make four kinds of wadokei clocks and nine types of mechanical dolls including the tea server.

A seventeen-page volume in the treatise entitled "Toys" provides a blueprint for the chahakobi ningyo. It was made entirely of wood, put together with wooden pins and powered by a spring of coiled baleen from a right whale. It explains how the baleen coil, when ratcheted up with a thick square key, powers the main propulsive cogwheel, which drives another, smaller one. On this and another wheel are attached long pins that carry the feet, which are seen to be "stepping" under the hakama skirts. This action also makes the

Seven kinds of wood are used to make the tea-serving doll, including cherry, bamboo and boxwood. The cogwheel is of multilayered quince wood and takes about three months to make. Due to seasonal humidity variations, it takes about a year to make the chahakobi doll parts and then three months for assembly. It has to be done when the humidity is right for wood expansion or the doll won't work. Traditionally, baleen from right whales was used to power the cogwheel (opposite).

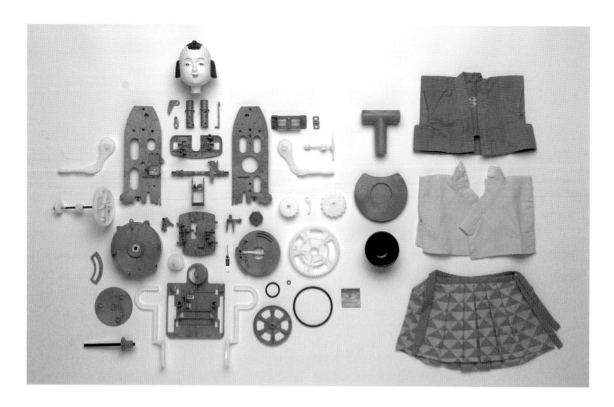

In 2002, Tokyo publisher Gakken Co. launched the first of its windup karakuri doll plastic kits based on Yorinao Hosokawa's tea server. The assembly manual carries explanations from his *Karakuri Zui* of 1796, and when correctly completed, the doll scoots across the floor holding a cup of tea until someone picks it up. When it is replaced, the doll does an about-face and returns to its starting point. "We think the products' appeal is not only learning about the clever abilities of these ancient craftsmen, but also how they instill Japanese with a sense of pride in their history," says Gakken editor Shigeru Kaneko.

head bob via strings running from the pins to the doll's neck.

The two-hundred-year-old book, which also details a tumbler doll with a mercury mechanism, and a windup doll that plays a drum and flute, comes complete with notes about the proper attire for the winsome tea server. With its eye for detail and concern for accuracy, Hosokawa's unique work has been highly valued in recent decades both for recording knowledge that would only have been orally transmitted from karakuri master to apprentice, and because it has allowed knowledge lost in Japan's rapid modernization to be reclaimed in new karakuri.

In the 1960s, when no functioning, original chahakobi dolls were left in Japan, Waseda University professor Shoji Tatsukawa and his students were able to create one after much trial and error with the help of Hosokawa's blueprints. The project received enthusiastic media coverage: the authoritative business daily *Nihon Keizai Shimbun* proclaimed the successful recreation of "a robot from the Edo Period."

The doll has also won over modern audiences, appearing in ads by a semiconductor maker and as a poster boy for events celebrating the four-hundredth anniversary of the founding of the Tokugawa shogunate in 2003.

In bringing the wheeled waiter into the twentieth century, Tatsukawa had the help not only of Hosokawa's blueprints but also of karakuri craftsman Shobe Tamaya VII. His son Shobe Tamaya IX continues to repair and produce his own karakuri puppets, including a South-Pointing Chariot with a *karako* Chinese acrobat puppet that was exhibited at Japan's 2005 Aichi Expo. In doing this, Tamaya is playing a key role in preserving the heritage underpinning Japan's world-leading robotics industry.

The Gadget Wizard

Hosokawa's quaint tea-carrying doll was the most famous of the *zashiki* karakuri, or "room" dolls. Designed as home entertainment for feudal Japanese, these took the technology of *butai karakuri*—the float puppets and karakuri displayed on stage at Takeda's shows—to another level, making them the true ancestors of today's humanoid machines. The tea server, for example, could move in a semi-autonomous manner, perform a human task and had human form, all attributes included in the several definitions of "robot."

"They shaped the way Japanese view robots," says Toyota National College of Technology Dean Yoshikazu Suematsu, a mechanical engineer and karakuri enthusiast whose collection includes a chahakobi doll made by Tamaya. "To put it simply, the difference between Mighty Atom and the Terminator shows the differences between how Japanese and Westerners view robots. Westerners tend to have this sense of alarm or wariness. Japanese are unique in the world for their strong love and affinity for robots."

The lifelike automata that began appearing in Europe in the seventeenth century, which were more technically sophisticated than anything produced in Japan, can also be seen as early robots, but Suematsu believes there is a difference. European automata, he says, were essentially attempts to reproduce human activities in machine form. The goal of Japanese karakuri was not realism but charm—art for its own sake rather than the advancement of scientific knowledge.

Hosokawa and the other Japanese karakuri masters of old, however, did not eschew pure science and technology in pursuing their desire to entertain, fascinate and surprise through their clockwork puppets. They were Japanese "Renaissance Men"—

The late Edo inventor Hisashige Tanaka has been called Japan's Edison. Not only did he create unsurpassed karakuri masterpieces, this son of a tortoiseshell craftsman invented innovative lighting and firefighting equipment, reproduced steam engines and founded the precursor of electronics giant Toshiba Corp.

Tanaka's *Yumihikidoji* archer (above) plucks arrows from its quiver, sights a target and lets fire. His long-lost *Mojikaki ningyo* calligrapher (below) was unearthed in the collection of U.S. magician Harry Kellar (1849-1922). Built without a single nail, it can write four Chinese characters (*kotobuki*, or longevity, seen here) with brush and ink. "No one would be able to make this doll today," says Osaka craftsman Susumu Higashino, who restored it. "This shows the world that Japanese had advanced technology in the Edo Period."

equally at home in the realm of art as in the fields of mathematics, engineering, chemistry and medicine—who devoured the Western knowledge that entered via the port of Nagasaki.

A giant among them was the late Edo inventor Hisashige Tanaka, born at the close of the eighteenth century as the son of a tortoiseshell craftsman in Kurume on Kyushu Island. At age nine he devised a trick ink-stone case that nobody could open; by twenty-one he was making his own karakuri and staging performances at a shrine. He went on to produce many practical inventions that earned him a reputation as Japan's Thomas Alva Edison, including an innovative oil lamp that used air pressure to ensure a stable oil supply and light ten times brighter than a candle. He also created his own karakuri dolls; like Hosokawa, he earned the nickname of Karakuri Giemon, which might be translated as the "Gadget Wizard."

Two of his greatest inventions exemplify the affinity between karakuri dolls and clocks. One was the extraordinarily complex *Mannendokei*, or Eternal Clock, a six-faced timepiece completed in 1851 and estimated to run for a year at a single winding. It is the pinnacle of wadokei technology and art.

Topped by a glass dome under which the current positions of the sun and moon were indicated over a map of Japan, the baroque instrument precisely told Western time, Japanese time, the seasonal divisions of the traditional calendar, the days of the week, Oriental zodiac dates and the phases of the moon. Cloisonné images of turtles and cranes adorned the clock's hexagonal base. When engineers refurbished the one-hundred-and-fifty-year-old device for display at the 2005 Aichi Expo, they were stunned at its mechanical intricacy.

But Tanaka's most famous karakuri invention is the *Yumihikidoji*, or Archer Doll. This zashiki karakuri consists of a tonsured boy archer in resplendent kimono seated on a platform with a quiver and a separate, small hanging target. The doll plucks an arrow from its quiver, carefully nocks it in the bowstring, sights the target while drawing the bow and fires. It may hit or miss the bullseye but invariably reacts with a facial expression that is the product of light and shadow playing on its porcelain-white face.

Six strings controlling its head make its face as expressive as the masks of the Noh theater, whose actors also employ the viewer's perspective to seemingly alter their expression. This

Arguably Tanaka's masterwork, the *Man-nendokei* or Eternal Clock (restored and in a nineteenth century flyer above) is a fascinating work of art. The positions of the sun and moon, as seen from Kyoto, are indicated in a glass dome at the top of this extraordinary timepiece.

conveyance of the puppet's "emotions" to the audience through subtle tilting and shadow allows joy, fear, anger and sadness to be expressed by an unchanging face; many other mechanical puppets including those on the festival floats employ the same technique. Suematsu notes that although Tanaka's Yumihikidoji performs the same movements regardless of whether it hits or misses, the viewer's own psychological state will influence him or her to interpret the doll's expression as joy in the former case and disappointment in the latter.

Tanaka's output was prodigious. He produced karakuri shows around the country and at age forty-eight opened a workshop showcasing mechanical inventions in Kyoto. Even in his seventies, after the Tokugawa regime fell and the Meiji Restoration launched Japan on a do-or-die modernization program, Karakuri Giemon was still inventing. At seventy-five, he founded a telegraphic machinery company, Tanaka Seizo-sho, and built Morse telegraphs and early telephones. He also experimented with bicycles and rickshaws.

The firm that Tanaka founded eventually evolved into Japanese electronics giant Toshiba, which today makes industrial robots as well as what it calls "life support partner" prototype robots for home use. With sophisticated image- and voice-recognition technology, the wheeled, cartoonish-looking ApriAlpha and ApriAttenda can carry out simple conversations, control household appliances and read out email. They may not be as thrilling to watch as Tanaka's archer doll, but they owe much to his legacy. In describing the heritage of Tanaka and other karakuri masters, Suematsu writes: "There is no doubt that Japan's mechanical puppet culture represented by these men contributed to industrial modernization after the opening of the country in the Meiji Period."

Tanaka's legacy—and the karakuri tradition—is anything but a fossilized art. At the 2005 World Karakuri Contest, judges chose the best new karakuri inventions in "classical" and "modern" divisions among more than six hundred entries from Japan and countries as far flung as Indonesia, the United States and Switzerland. The one million yen Grand Prix award went to the maker of a doll on horseback that fires arrows at targets while moving, a karakuri version of traditional Japanese *yabusame* mounted archery.

But the event featured an even greater marvel, a restored Tanaka karakuri masterpiece appearing for the first time in one hundred and fifty years. It was an unknown work called the *Moji-kaki-ningyo*, or Writing Doll, a female karakuri clad in a kimono and kneeling on an ornate, twelve-inch-tall stand in front of a small writing board. She clasps a brush in her right hand and is operated by a complex windup mechanism. Incorporating intricate cams and gears in the base, the doll is believed to have been created by Tanaka in the 1840s, and was brought back from the U.S. in 2004, but how it left Japan is a mystery, leading experts to dub it the "Phantom Karakuri."

Another intriguing question was what the Moji-kaki doll would write after being restored to its original state. Scholars were so impressed with the sophistication of the mechanism that some suggested the doll would rewrite the history of Edo-period technology. What did the Gadget Wizard's calligrapher actually write with her brush? Naturally, the Chinese character *kotobuki* (寿), which means "longevity."

CHAPTER 2   The Buddha Robot

Isn't it sad to be content with merely making robots as slaves?
—Makoto Nishimura

Japan's first modern robot, built in 1928, was a direct descendant of the medieval karakuri dolls, created to awe and entertain. Rather than a robot doll, however, it was more a robot Buddha—a giant golden man that could move its upper body, change its facial expression and write Chinese characters while sitting at an ornate altar-like desk.

This unique marvel may have been years ahead of its time technologically, but the man who built it was no clockmaker, mechanical engineer or inventor. Makoto Nishimura was a biologist and botanist whose life took him from the mountains of rural Nagano Prefecture to the halls of Columbia University in New York. He could draw, paint and sculpt. He was also an impeccable dandy in dress with a lean, aristocratic face.

Born at the foot of the Japan Alps in the castle town of Matsumoto in 1883, he went to teachers' college in Hiroshima and then worked at schools in Kyoto and Japan's puppet state of Manchukuo in Manchuria. After a spell studying in the United States, by 1925 he was back in Japan on what was then a frontier island, Hokkaido, doing research on the indigenous Ainu people and teaching biology at Hokkaido Imperial University. It was there that he wrote a 1927 essay, entitled "The Pacific Fifty Years From Now," for an Osaka newspaper which so impressed the publisher that he offered Nishimura a job as a columnist and consultant. Though he was a well-known scholar at the time, Nishimura quit his secure teaching post in Japan's colonial north

A scaled-down replica of the Buddha-like Gakutensoku, acknowledged as Japan's first robot, stares at visitors to the Osaka Science Museum. Created by Makoto Nishimura in 1928, its lifelike movements stunned Japanese at an exhibition in Kyoto.

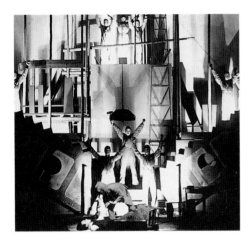

Gakutensoku inventor Nishimura was influenced by Karel Čapek's groundbreaking 1921 play *R.U.R. (Rossum's Universal Robots)*, seen here in its second Japan production at Tokyo's Tsukiji Little Theater in 1926. Čapek gave the world the word "robot" and helped propagate the stereotype of technology running amuck. Designed by Kenkichi Yoshida, the set of *R.U.R.* featured massive doors that closed like two saw blades engaging one another.

to become a newspaperman.

Nishimura was also an uncompromising environmentalist. In erecting a house, he decided against cutting down a large tree on his plot. He built around it instead, so the living room surrounded the tree's trunk and its branches went up through the roof. Philosophically, he was influenced by the spiritual doctrine of *ban-butsudokon*, which holds that all creation arises from the same source. And as remarkable as he was for his time, Nishimura might have been forgotten by history were it not for a story written half a world away. A story about the rise of machines.

"Then call them 'Robots'"

What exactly is a robot? The word is difficult to define. The *Oxford English Dictionary* says a robot is "a machine (sometimes resembling a human being in appearance) designed to function in place of a living agent, esp. one which carries out a variety of tasks automatically or with a minimum of external impulse." Japan's authoritative *Kojien* dictionary describes a robot thus: "a complicated and elaborate manmade automaton, an artificial person or cyborg; in general, a machine or aparatus for work or control that can be made to perform automatically."

To qualify as a robot, it seems a machine's function is more important than its appearance—robots do not need to be in human form to be robots, but generally they do human work, often jobs deemed too dangerous, difficult or monotonous for us. Curiously, people can also be robots. When someone acts in a mechanical fashion, without showing emotion or sensitivity or obeying orders automatically, we might call them a robot. The word thus embraces the idea of machines behaving like humans as well as humans behaving like machines.

"Robot" was introduced by Karel Čapek, one of the most important Czech writers of the twentieth century and a prominent dissident amid the interwar rise of fascism in Europe. Like George Orwell, he was a mainstream author, a journalist and a poet who sometimes exploited science fiction motifs to express his views about technology and authoritarianism. Čapek was on summer vacation in a Slovkian spa town when he got the idea for a play about synthetic factory workers rising up against their human creators. He approached his artist brother Josef, who was holidaying with him, and said he had thought of calling

Works by the visionary Čapek remain popular in Japan today, as seen by this poster for a May 2000 production of *R.U.R.* staged by the Haiyuza Theatre Company in Tokyo.

his artificial laborers "*labori*," but this sounded too bookish. Josef, feverishly painting at the time, simply muttered through a brush in his mouth: "Then call them 'Robots.'"

The word was derived from the archaic Czech "robota" meaning obligatory work performed by medieval serfs, or "robotniks." Čapek's play *R.U.R. (Rossum's Universal Robots)* used the familiar Frankenstein motif of human hubris producing artificial beings that become uncontrollable and violent.

The company in the play, R.U.R., mass-produces robots—made of flesh and blood, not metal and electricity—that are the ultimate in efficient labor. Their inventor, old Rossum, "wanted to somehow scientifically dethrone God," as R.U.R. head Domin tells an idealistic visitor to the factory bent on liberating the robots. "For him the question was just to prove that God is unnecessary. So he resolved to create a human being just like us, down to the last hair."

But the utopian society that is sustained by worldwide robot labor has a dark side: people become sterile, robots show signs of having souls and are even being used as soldiers. When they revolt, killing all but one of the factory's managers, their leader Radius proclaims: "The age of mankind is over. A new world has begun! The rule of Robots!" Without old Rossum's secret formula, though, the robots cannot replicate themselves. In the end, the natural order is restored when two robots fall in love and are christened Adam and Eve by the sole surviving human.

Čapek later wrote that the play's aim was not to condemn technology per se but to question the idea of man becoming slave to the ascendant machine, the triumphant factory. He saw his dark play as "a comedy, partly of science, partly of truth." "We are in the grip of industrialism," he added. "A product of the human brain has at last escaped from the control of human hands."

A Word, a Phenomenon

R.U.R. was an instant success when it was staged in Prague in 1921, and earned Čapek international renown with productions in New York, London and Paris soon after. The play also fueled the growth of American pulp science fiction magazines like Hugo Gernsback's *Amazing Stories* and their negative portrayal of robots.

Čapek's artificial workers also made a deep impression on Japanese, including Nishimura. In July 1924 *R.U.R.* came to Tokyo's Tsukiji Little Theater, the country's first playhouse for modern Western drama. It was titled *Jinzo Ningen* ("Artificial Human"), and was the venue's first work staged in its entirety. Kihachi Kitamura, a translator who saw the production, left this reflection: "I think the author's intent was to show people controlling the ultimate in science, yet not losing human love—that's where the future of humanity lies." The interest in *R.U.R.* led to a repeat performance at the theater two years later. The ideas in the play also helped prime Japan for its first "robot boom" in the late nineteen twenties.

Many Japanese found the concept of an artificially created human intriguing instead of terrifying, and jinzo ningen became a catchword. Čapek's influence was soon seen in Japanese literature; at first writers tended to imagine robots like Rossum's organic creations of flesh and blood, and not mechanical drones.

In August 1924, for example, a story by Shiro Kunieda appeared in *Shosetsu Kodan Poketto* magazine that was entitled "*Tantei*

The Gakutensoku replica and the Chinese characters for "Osaka." Nishimura created his pneumatic man to impress and delight at a time when robots were seen as potential replacements for human workers.

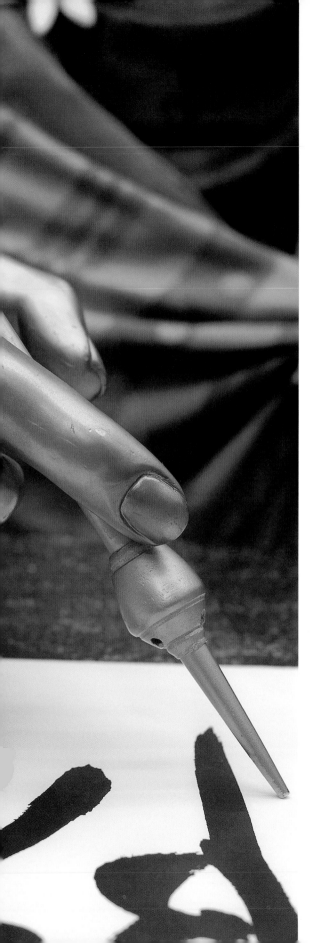

Kibun Ningen Seizo" (A Detective's Bizarre Tale of the Manufactured Man). It concerns a detective's discovery of a secret experiment involving an American doctor and a Japanese nurse to create, by means of using various medicines and substances, an eight-foot artificial man. Of course, the only part of the giant they cannot concoct is the heart, so they set about relieving unsuspecting people of theirs. Robot historian Haruki Inoue sees this as the first Japanese work influenced by *R.U.R.*, although basically a retelling of it with the American in Rossum's role.

Japanese scientists were also interested in Čapek's vision. In 1926, monthly science magazine *Kagaku Gaho* published a special titled "A Scientific View of the World in 100 Years." Hisomu Nagai, a professor of physiology at Tokyo Imperial University who was also a proponent of eugenics, contributed an article headlined "*Jinzo Ningen wa Kano ka*" (Is an Artificial Person Possible?) that was illustrated with stage scenes from *R.U.R.*

The term "jinzo ningen" eventually became the catchier *robotto* (robot), which only made it into a dictionary of new words in 1928. But that was the year everyone had something else to talk about.

The Machine That Smiled

The formal ascension to the Chysanthemum Throne by the youthful Emperor Hirohito in 1928 called for huge public celebrations. When biologist turned newspaperman Nishimura learned that his employer, the *Osaka Mainichi Shimbun*, was planning to enter an exhibit at one such celebration—a fair in Kyoto—he proposed building an "artificial man."

In contrast to the prevailing *R.U.R.*-influenced concept of an organic automaton, Nishimura set about planning a mechanical being. Since he wanted it to be lifelike, though, the automaton's body would need to move smoothly. One of his talents was playing the *shakuhachi*, the Japanese bamboo flute, and it was during a practice session that he conceived of the principal mechanism for motion. He noticed how stopping the instrument's holes with his fingers changed the air flow and the pitch, and hit upon the idea of using compressed air.

With the aid of several assistants, the biologist designed and built a magnificent seven-foot, eight-inch-tall humanoid that could open and close its eyes, smile and write Chinese charac-

Botanist Nishimura, seen here designing Gakutensoku's giant hand, knew more about living creatures than how to build a robot. But he devised a system of rubber tubes to pipe pressurized air through his robot's torso and arms, allowing it to move smoothly and exhibit a range of realistic facial expressions.

ters all via air pressure, gears and pulleys. He gave it the name *Gakutensoku*, which means "learning from natural law." It was designed to be more akin to a living creature than a mechanical tool—and a small cosmos flower on its breast, along with engravings of birds, plants and snakes on the desk at which it sat, emphasized the theme.

Gakutensoku consisted of a steel and wood frame covered with rubber skin created from molds that Nishimura sculpted himself. Its face was a combination of different racial features meant to symbolize universal solidarity. Besides his shakuhachi, Nishimura was inspired by everything from human reflexes to snake locomotion. The key mechanism, however, was the air pressure regulator to the rubber tubes inside the robot's arms and torso. Air pressure was also used to make its cheeks puff in and out as if breathing, and an array of springs and gears in its large head imparted facial expressiveness. It was thus designed in every way to seem "alive."

Visitors to the exhibition hall in Kyoto were treated to something like an audience with the Wizard of Oz. The show began with an empty stage reminiscent of a classical Greek temple. An enormous artificial bird on a circular canopy over the stage moved its head about and sang. Soon a screen parted, solemn music began to play and a giant torso appeared from within, seated at an ornate altar and wielding a pen on the end of an arrow shaft in one hand and a mace in the other.

The crown of laurel leaves and the toga-like robes, along with the other rather intimidating accoutrements, gave it the aura of a Roman personification of justice (if one were to disregard the figure's pointy ears). But Japanese who saw it would have likely imagined a statue of Buddha—a familiar feature in every community temple—had come to life. Bathed in green and red light, the golden man began to slowly move. Stunned visitors watched as it held up its mace to its forehead. When the weapon lit up, Gakutensoku opened its enormous eyes as if struck by a bolt of lightning, looked up to the sky, then smiled. It began to write Chinese characters with its arrow pen.

Novelist Hiroshi Aramata describes the audience reaction thus: "It started to write characters smoothly in a flowing hand. As if to express the agony of creation, it slowly shook his head from left to right. The movement was so natural it didn't look

Gakutensoku's face was designed to reflect the characteristics of several races. Its name means "learning from natural law."

An archival photo of the original Gakutensoku, which was nearly eight feet tall and could write Chinese characters like the karakuri automata of old Japan. Some Japanese were so impressed they offered prayers to it.

like it was a machine. Unconsciously, the spectators began naturally imitating this movement, shaking their heads from left to right. This was funny because the humans looked like they were being controlled by the robot like marionettes."

More Mechanical Men

In retrospect, it is remarkable that Japan's first steps into what would become the world's top robot-producing country—the Robot Kingdom—were so different than the rest of the world. Its first robot, Gakutensoku, was fundamentally unlike its automaton contemporaries in the United States and Britain. They were mechanical in appearance and utilitarian, closer, in purpose, to the robots of Čapek's drama. Westinghouse Electric and Manufacturing Co.'s 1927 robot Televox consisted of telephone switching equipment behind a crude human-shaped front like a cardboard cutout. It could respond to human phone commands issued by producing different pitches on a whistle and carry out tasks such as checking water levels or activating electrical appliances. Televox caused a national sensation as it toured the country, and despite the machine's lack of genuine anthropomorphism, the *New York Times* hailed its birth with the headline: "Science Produces the 'Electrical Man.'"

England's Eric, meanwhile, looked more like a real robot, but, with a metallic shell like a knight's suit of armor, a decidedly European one. Complete with breastplate emblazoned with the letters R.U.R. it was unveiled at an engineering exhibition in 1928. It used batteries, pulleys and an electric motor to rise from its seat and look around, allowing an operator to make it "speak" to onlookers through a radio within; sparks spewed from its mouth when it did so.

Both Televox and Eric were clearly designed as artificial laborers and tools—in fact, three Televox units were actually employed by the U.S. War Department to monitor reservoirs in Washington, D.C. Gakutensoku's job, however, was simply to be. It did not represent a new class of synthetic slave workers like Rossum's robots. Like a giant karakuri doll, Nishimura's robot was created in the image of man, for the pleasure of man. As Aramata notes, Nishimura designed Gakutensoku as "an attempt to set aesthetic robots free from being slaves to industry."

The Japanese humanoid also came with a message from

Kyoto sculptor Masahito Iwano has made a replica of Yasutaro Mitsui's robot, seen on the opposite page. "There's something really romantic and perhaps artistic about these old robots," says Iwano, an admirer of Makoto Nishimura. "They were one of the many ways people in the past imagined the future. When Japanese heard about Čapek's *R.U.R.*, Televox and others, they associated them in their minds with human beings."

Nishimura; an attached shield bore the following motto in Japanese and English: "Man's maker is God. Man findeth God in his work, Without God civilization becometh a curse." For a *Sunday Mainichi* magazine special edition about the exhibition, Nishimura contributed an article introducing his creation in which he also referred to Čapek's *R.U.R.*, writing: "If one considers humans as the children of nature, artificial humans created by the hand of man are thus nature's grandchildren."

Instead of a servant or manual laborer, Gakutensoku was designed to be an "ideal man" expressing the marriage of art and science, says Kenichi Kato, chief curator of the Osaka Science Museum, where a non-functioning replica of the golden robot is on display. "Nishimura was a botanist so he believed the world should be governed according to the rules of nature. That's the meaning of Gakutensoku," says Kato. "He borrowed the robot idea from Čapek, but revised it. He thought robots should be ideal humans."

The First Boom

The year 1928 was the beginning of an exceptional time for robots in Japan. Along with Gakutensoku, reports of foreign automatons in the press began a robot boom that continued into the 1930s. In 1929, Fritz Lang's *Metropolis*, which featured the female robot Maria, opened in Japan and proved a wild success. The following year saw an exhibition entitled "Tokyo in the Year 1990" at the Matsuzakaya department store in the capital's Ueno area that featured a lifelike German robot dressed in a uniform and named Remarque, after *All Quiet on the Western Front* author Erich Maria Remarque. Archival images show it saluting a large crowd with a broad smile and doffing its peaked cap in front of two ladies in kimono.

The media-fueled interest in robots in Japan peaked in 1931, which saw a flurry of comic strips, political cartoons, songs and ads for everything from insecticides to radio sets featuring robots. Often they were portrayed as farcial or friendly tin men that could grace the cover of a children's textbook or ad for a department store. The decade also saw the introduction of Gajo Sakamoto's popular cartoon robot character Tanku Tankuro, a cannonball-shaped cyborg that could pull all manner of appendages or tools from orifices in his body and change into a plane

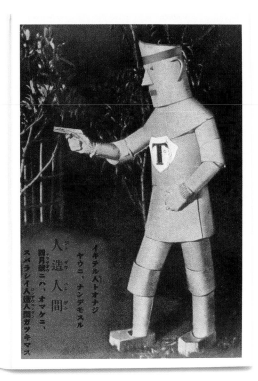

Haruki Inoue's 1993 book *Nihon Robotto Soseiki 1920–1938* explores Japan's first robot boom that followed the staging of *R.U.R.* in the 1920s. Like Nishimura, Yasutaro Mitsui built his own giant steel humanoid, and posed with it for a formal portrait in the early 1930s (above left). Robots were often featured in children's magazines of the time, such as this bonus picture postcard (above right) from the March 1932 edition of *Yonen Kurabu*. The caption reads: "The Jinzo Ningen—It can do anything, just like a living person."

or tank. Juza Unno, a government-employed electrical engineer, became revered as the father of Japanese science fiction in part for penning stories in the 1930s about Japanese robots fighting the Chinese. Inspired by the work of Nikola Tesla, Unno in turn influenced comic genius Osamu Tezuka, creator of Mighty Atom, Japan's foremost fictional robot.

Disappearing Act

With all the attention, it is easy to see why Nishimura believed Japan's Robot Kingdom was at hand. "Anyone who reads Karel Čapek's play *R.U.R.* would think that it's just a fantasy," he wrote. "Of course, the way things are now, it must be just a fantasy to think about educating robots. But actually, the era of making robots is beginning." Less than a decade after Gakutensoku was made, however, Japan was at war with China and the robot boom faded. Nishimura's dreams were put on hold, and he worked as a schoolteacher and in social welfare until his death in 1956 at age 73.

But the fate of Gakutensoku is shrouded in mystery. After the Kyoto exhibition of 1928, it departed on a tour of Germany—where it disappeared, apparently without a trace. Yet Nishimura's unique machine didn't vanish entirely from the Japanese consciousness. Besides the mockup at the Osaka Science Museum and citations in books and websites, author Aramata featured Gakutensoku in his 1980s fantasy novel series *Teito Monogatari*, which was made into animated and live-action films. In the story, when 1920s Tokyo is threatened by an evil magician, Nishimura and a heroic Gakutensoku take on a nasty gang of the sorcerer's demons blocking construction of the subway.

And, if only in name, Nishimura's dream is also on an endless journey through space, circling the sun in perpetuity. In 1995, astronomer Takao Kobayashi named an asteroid hundreds of millions of miles from earth in honor of Japan's first robot, "Gakutensoku (9786)." If the golden man were still around, doubtless he would look up to the heavens and smile.

CHAPTER 3

100,000 Horsepower Dreams

Beyond the sky, la la la
Reaching for the stars
Go, Atom
As far as your jets can take you
With a warm heart, la la la
Child of science
100,000 horsepower
Mighty Atom!

—original *Mighty Atom* TV series theme song lyrics

On a dreary April morning, a crowd had braved the rain to gather outside the train station in Takadanobaba, a Tokyo neighborhood of smoky bars and yakitori joints popular with students from nearby Waseda University. Many were dressed in colorful pointy caps, fuzzy white wigs and red boots. A brass band led a parade of hundreds of the costumed revelers singing "Happy Birthday." They were celebrating the Birthday Eve of Mighty Atom, a hero to people around the globe. The next day, April 7, 2003, was—according to the original 1950s comic series—when Japan's best-loved fictional robot was born in a Takadanobaba laboratory.

That day, celebrations reached fever pitch. Hordes of fans pressed into a museum on the outskirts of Osaka to watch a reenactment of Atom coming alive on the lab table. The Japan Mint announced special coin sets to commemorate his birth. The city of Niiza, home to a studio producing an Atom animated series, granted the robot honorary citizenship in a ceremony with a man in an Atom costume. The Ministry of Economy, Trade and Industry opened a public exhibition in its lobby showcasing

robots designed to help people—including Wakamaru, a robot watchdog and a robot vacuum cleaner. And in the heart of Nihonbashi, Tokyo's oldest shopping district, the Takashimaya department store was displaying a four-inch-tall Mighty Atom figurine covered with diamonds, emeralds and rubies worth nearly ¥100 million.

The effect of this small, pointy-headed superhero on Japanese attitudes toward humanoid robots is incalculable. While Atom is as iconic in Japan as Mickey Mouse is in the United States, he is also far more than just a cartoon character. His fans range from preschoolers to scientists. Atom embodies a deeply ingrained postwar vision of pacifism and technology, representing the wellspring of an almost universal agreement among theorists, researchers and engineers that robots can not only be friends with human beings but even be, perhaps, the country's salvation. That philosophy was born out of the death and destruction of a fire-bombed Osaka during World War II in the mind of a young medical student who would one day become Japan's "manga god."

The God of Comics

Tetsuwan Atomu (Mighty Atom, also known as Astro Boy), was the brainchild of Osamu Tezuka. Born in 1928 in Osaka, Tezuka grew up as the eldest of three children in a well-off, liberal family in the nearby satellite city of Takarazuka. His father was a comics, movie and photography buff. His mother, a descendant of legendary Japanese ninja Hattori Hanzo, patronized the local Takarazuka Revue, a famed all-female, Western-style musical company that still has an enormous national following; two of its starlets lived next to the Tezukas.

Short and thin, Tezuka was continually bullied at school because he wore glasses and had unkempt hair that earned him the nickname "fuzzy head." But the insects and wild animals that inhabited his neighborhood of Gotenyama provided a world of escape, and his imagination was fueled by his mother's stories. She also took him to see the Takarazuka musicals and the cinema—influences that would manifest themselves later on.

Tezuka found food for his voracious imagination everywhere, and even the Western meals he ate at a fancy restaurant in the Takarazuka opera house as a child would turn up in his comics. He also received encouragement from his father, who had drawn

Growing up in a well-off family in Takarazuka north of Osaka, Tezuka, pictured here at age four, was influenced by his father's love of comics and animated shorts, taking to drawing at a young age.

his own cartoons in his youth. The elder Tezuka would screen animated shorts at home, such as *Popeye* and *Mickey Mouse* in English, and he subscribed to the monthly boys' comic *Shonen Club* so the family could read children's strips such as the immensely popular *Norakuro* (Blackie the Stray) by cartoonist Suiho Tagawa.

Tagawa's tale of an orphaned puppy who joins the Japanese Imperial Army in pursuit of his dream of becoming a famous general and, despite his clumsiness, manages to rise through the ranks, was a hit, spawning Disneyesque merchandising. Like other periodicals of the time, *Shonen Club* toed the line of Japanese militarism, even teaching readers how to throw hand grenades in its July 1945 edition, though the Imperial Army considered the bumbling Norakuro bad publicity. Tagawa's battle-scarred hound, the only black dog in a regiment of white dogs, raised issues of isolation and identity that later colored Tezuka's own oeuvre.

Osamu began to draw at an early age, beginning with cartoons of his teachers and neighbors. By the fifth grade, he was sketching comics of people on Mars. An avid reader, in 1939 Tezuka borrowed a friend's insect encyclopedia and was inspired by a ground beetle in it known as the *osamushi* in Japanese. As a homage to the beetle, he decided to add the Chinese character 虫 meaning "insect" to his first name to create his celebrated penname. In middle school, he became engrossed with drawing original manga with titles such as "A Record of the Insect Battle Front" and "The Old Man Detective."

As a teenager, Tezuka worked in a factory like many other students who were filling manpower shortages from the military call-up for the war. There he experienced the catastrophic death and destruction of the Osaka firebombings; it left a deep impression on the young man, far surpassing the pain from the severe beatings he received from teachers for drawing manga.

At the war's end, he entered Osaka University's Faculty of Medicine, but even the arduous life of a medical student failed to dampen his enthusiasm for drawing. In 1946, at the age of eighteen, he published his first serialized comic in the *Shokoku-min Shimbun* newspaper. It was entitled "The Diary of Ma-chan," a four-frame gag strip about the hi-jinks of preschooler Ma-chan and his buddy Ton-chan. It proved so popular that wooden Ma-chan dolls were produced, the first of what were to be innumer-

A workaholic throughout his prolific career as an animator and cartoonist, Tezuka penned some 150,000 manga pages before he died in 1989. He pioneered the graphic novel in Japan and a range of genres, not the least of which was the robot as hero.

able Tezuka merchandise spinoffs. The following year saw the publication of Tezuka's first hardcover manga, *Shin Takarajima* (*New Treasure Island*), a massive success that was based on a script by veteran Osaka cartoonist Shichima Sakai.

The simple yarn about a boy, a treasure map and pirates was immensely more graphically sophisticated than Ma-chan and gave a different look to the manga genre. At the time, foreign movies were flooding Japan following the ending of wartime restrictions. Tezuka later wrote: "Why are American movies so different from Japanese ones? How can I draw comics that make people laugh, cry and be moved?" With *New Treasure Island*, Tezuka tapped the energy of the motion picture through quick-cut frames, dramatic close-ups and long shots. The opening sequence showed boy hero Pete racing his car to the harbor in cinematic style. The manga sold over four hundred thousand copies and went through several editions.

Tezuka was all of twenty.

Atomic-powered Pinocchio

Flush with success but nevertheless determined to complete his medical degree, Tezuka traveled to Tokyo, where he showed his work to established manga artists. His genius was quickly recognized, and jobs began to flow in. Soon, major Tezuka narratives like *Jungle Emperor*, the classic story of how an orphaned lion cub claims his heritage as ruler of the animal kingdom (broadcast overseas as *Kimba the White Lion,*) were being serialized in popular comic magazines.

It was here—in one of these publications, *Shonen*—that Japan's most famous robot first appeared. He debuted in a series called *Atomu Taishi* (Ambassador Atom) that began in April 1951, and he played a role that early on outlined his creator's vision of peace and humanitarianism. When aliens who look like ordinary humans come to twenty-first century Earth after the destruction of their own planet, food shortages cause conflict between the two species. Atom was introduced in the fourth installment of the series as a performer in a machine circus.

Since Tezuka liked to insert himself at the beginning of his Atom stories in Alfred Hitchcock-style preludes, we can trace the history of Atom's development. Before one story published in 1952, Tezuka relates how his editors asked him to make the

© Tezuka Productions

Atom first appeared in this April 1951 series called *Atomu Taishi* (literally, Ambassador Atom) as a performer in a machine circus. As a peace envoy, he plays a decisive role in averting a nuclear war between people on earth and a race of aliens. The robot boy would continue to play the part of mediator between opposing groups in many Atom tales.

humanlike robot the star of his own comic, then isolated him in a hotel room to get the job done. Despite writing "under pressure" and without having "worked out everything in my mind yet," Tezuka produced something very special. Of course, it's unlikely the editors had any idea of the scale of what they would set in motion: the hotel-room drawings were to lead to a wildly successful eighteen-year comic series running to some five thousand pages and Japan's first-ever animated television series in the 1960s. It was later to be broadcast in over fifty countries.

The story of *Tetsuwan Atomu* features an elaborate backstory for the robot. On April 7, 2003, Atom is created by grief-stricken scientist Dr. Tenma of the Science Ministry following the death of his son Tobio in a traffic accident. In his famous birth scene, Atom rises from a table in Tenma's Takadanobaba lab much like film depictions of Frankenstein's monster. But Tenma, desperate to replace Tobio, treats Atom like a person, not a synthetic creature, and his invention becomes more human with each passing day.

Under Tenma's care, the doe-eyed robot child demonstrates elements of what today's computer scientists and engineers might term emergent behavior from machine learning. As the narrator says of the "resurrected" Tobio: "Even his eyes became more and more humanlike in the way they sparkled. One month later, Tobio smiled. For the first time, his 'pleasure circuits' had created a happy expression. And his smile was as pure as that of an angel." Ironically, Atom first smiles when Tenma gives him a present—a toy robot.

Atom's inability to grow, however, comes as a stunning revelation to Tenma, shattering the illusion of Tobio's return from the dead. The scientist then violently rejects his own creation: "This thing's just a doll! It's not my son! Get out of here, you freak!" Atom is sold into the robot circus, where his humanlike appearance and seven awesome powers amaze the crowd. His famous "100,000 horsepower" strength would be about double the capacity of the *Titanic's* engines. Powered by atomic fusion, Atom can also fly at speeds of Mach five with the rocket jets in his arms and legs, shoot bullets with machine guns in his backside, see in the dark with his searchlight eyes, hear things miles away with ears one-thousand times more sensitive than humans, understand sixty languages, and see into the hearts of people to discern whether they are good or bad. Fortunately, Atom is rescued from the circus by the long-nosed, kindhearted Dr. Ochanomizu, who later creates robot parents and siblings for him.

The Machine as Hero

In *Ambassador Atom*, Tenma's robot is sent to the aliens' spaceship as an ambassador of peace before the aliens and humans annihilate each other with hydrogen bombs. Offering himself as a sacrifice by literally giving them his own head in trust, Atom earns the aliens' confidence, sealing a concord between the races and avoiding all-out war.

Atom played a minor role in the comic, but one central to his lasting significance in Japan as a robot that fought for peace and a poster child for benevolent technology. A machine rejected by his creator for not being human enough, Atom had to suffer the indignity of the robot circus before he saves the world as an intermediary between humans and aliens. In offering to sacrifice his own life to broker peace, Atom proves nobler than his belligerent human creators and thus redeems the essential flaw of being a mechanical creation. With this bold vision of a peacemaking technology that could transcend the worst of human nature, Tezuka presented a new kind of hero to Japanese children of the 1950s—and set the stage for future robots and roboticists.

Atom was not the first robot to appear in Japanese manga, even as a main character. Tezuka may have been influenced by Gajo Sakamoto's popular *Tanku Tankuro*, which remains in print seventy years later. He was definitely affected by the dark vision of human hubris and technology of Čapek's *R.U.R.*, which he read at age ten. But Atom was the first robot hero portrayed sympathetically and in stories that dealt with such serious dramatic themes as discrimination, environmental destruction, fascism, the causes of human conflict and humanity's future.

Tezuka's own views on technology, science and robots are still being debated. Deeply disturbed by war, he still held faith in modernization and science yet continually questioned whether they could redeem human depravity. "Atom is given life as something which stands between absolute truth and man," writes Keio

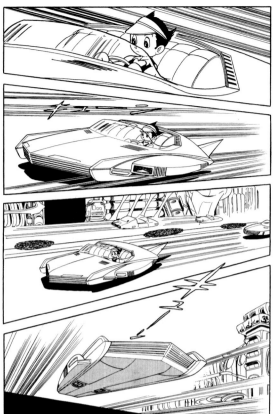

When Tezuka published *Shin Takarajima* (New Treasure Island) in 1947, readers were stunned by its cinematic style. The comic master used the technique again in this retelling of Atom's creation, in which he first introduces himself and his early views on robots in the future (above, far left) before the story begins with Tobio scooting along in a flying car toward his doom. Dr. Tenma uses all the powers of his science ministry to replace his fatally struck son with a robot boy, Atom (right).

University art history professor Yukio Kondo. "Carrying the imperfectness and contradictions of man in him, Atom asks the following question in eternal solitude: 'Why are men so foolish?'"

Frederik L. Schodt, a manga and Japanese robot historian who was Tezuka's friend and personal translator for years, believes the deeper themes present in the Atom manga reflect Tezuka's experience of war, medicine and reading Western literary classics. "If you read the stories carefully, Tezuka displays a strong skepticism about the benefits of technology," he says. "In fact, throughout his works, such as *Atom*, *Phoenix*, and others, he displays a profound awareness of the weakness of man, and of the danger that man will misuse technology."

From "Little Boy" to Mechanical Moralist

Tezuka would spend almost half his manga career drawing the little robot, wowing readers with gripping, richly imagined stories about masked villains in shadowy conspiracies, brilliant but short-

sighted scientists creating beings that run amok, wars between robots and humans and invading extraterrestrials. Robots are central to the Atom story and vividly illustrate the way Tezuka viewed technology. Like many Japanese who survived the U.S. atomic and fire-bombings of cities during World War II, he knew the awesome and horrific power of science all too well.

The comics, however, reflect Tezuka's view of technology itself as value-neutral, knowledge that could be used for either good or evil. The deciding factor, as he saw it, was human intent. The story in which Dr. Tenma uses all the computer resources of the government and the "world's most advanced technology" to "breathe life" into his robot son also sees the scientist—driven mad by grief over the loss of his son—launch a campaign to exterminate the aliens. As head of the secret police agency, Tenma deploys his ruthless "red shirt brigades" to kill them off, a plan Atom denounces in an impassioned rejection of his creator: "Anyone who does something *that* bad can't be my father!"

Junji Ito, an art critic and curator, has even argued that Atom

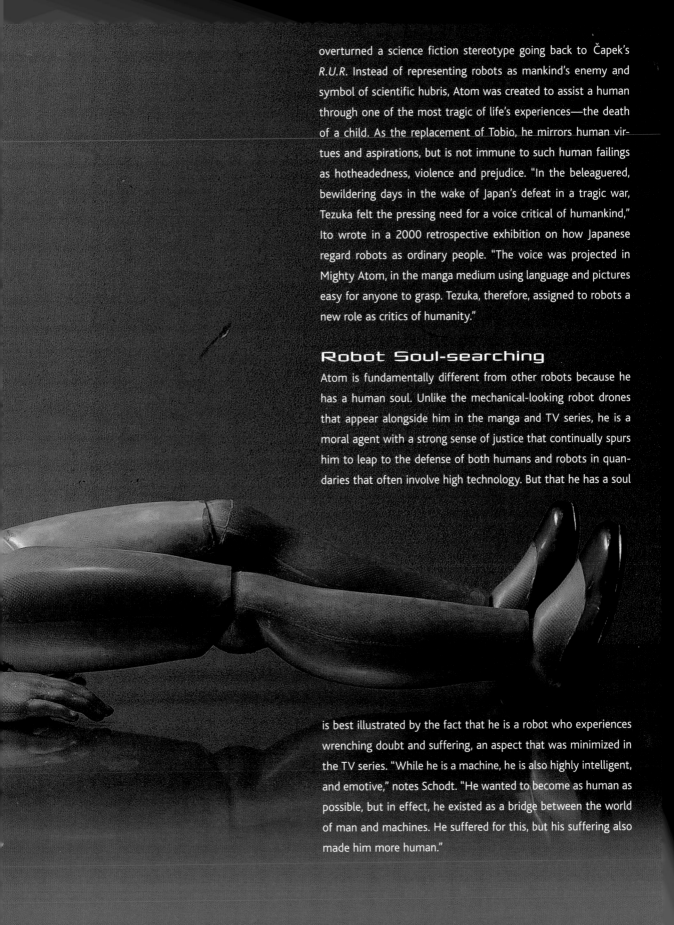

overturned a science fiction stereotype going back to Čapek's *R.U.R.* Instead of representing robots as mankind's enemy and symbol of scientific hubris, Atom was created to assist a human through one of the most tragic of life's experiences—the death of a child. As the replacement of Tobio, he mirrors human virtues and aspirations, but is not immune to such human failings as hotheadedness, violence and prejudice. "In the beleaguered, bewildering days in the wake of Japan's defeat in a tragic war, Tezuka felt the pressing need for a voice critical of humankind," Ito wrote in a 2000 retrospective exhibition on how Japanese regard robots as ordinary people. "The voice was projected in Mighty Atom, in the manga medium using language and pictures easy for anyone to grasp. Tezuka, therefore, assigned to robots a new role as critics of humanity."

Robot Soul-searching

Atom is fundamentally different from other robots because he has a human soul. Unlike the mechanical-looking robot drones that appear alongside him in the manga and TV series, he is a moral agent with a strong sense of justice that continually spurs him to leap to the defense of both humans and robots in quandaries that often involve high technology. But that he has a soul

is best illustrated by the fact that he is a robot who experiences wrenching doubt and suffering, an aspect that was minimized in the TV series. "While he is a machine, he is also highly intelligent, and emotive," notes Schodt. "He wanted to become as human as possible, but in effect, he existed as a bridge between the world of man and machines. He suffered for this, but his suffering also made him more human."

A perennial hero to countless Japanese children, Atom is a manga character that has also inspired serious art. Tokyo machinery sculptor Hiroshi Araki's 1993 "Astroboy" is an extraordinary work of carbon and aramid fiber, about three and a half feet long, that was conceived as a kind of memorial upon Tezuka's death. "Even though Atom is a manmade thing, I ultimately see him as an angel," says Araki, whose works include fanciful anthropomorphic cars, stereo speakers and cameras. "Just as angels mediate between heaven and the world of humans, Atom mediates between machines and people."

Atom is not immune to torn loyalties. Often more a willful, moody child than machine superhero, he openly probes his allegiances during a robot uprising in 1952's "Frankenstein" episode. He even joins a robot rebellion sparked by human abuses in 1965's "The Blue Knight." Azusa Nakajima, a novelist and manga critic, has described Atom as "a mind in agony trying to establish its ego."

Identity was an important theme throughout Tezuka's work. His generation was afflicted by feelings of dislocation, rejection, loss and betrayal following the collapse of the nationalist and militarist state ideology. Contemporary artist Takashi Murakami, whose postmodern sculptures and paintings are heavily influenced by manga and anime, has interpreted postwar popular culture as the portrayal of eternal childhood born in the annihilating terror of nuclear holocaust.

"Whatever true intentions underlie 'Little Boy,' the nickname for Hiroshima's atomic bomb, we Japanese are truly, deeply, pampered children," Murakami writes in *Little Boy: The Arts of Japan's Exploding Subculture*, a book accompanying a popular 2005 exhibition. "It's the denouement of a culture, nourished by trauma, snugly raised in the incubator of a society gone slack."

The subtext of Japan as war orphan was present in the very first serialized episode of the *Tetsuwan Atomu* manga. In "Gas People," published as the seven-year U.S. military occupation of Japan was ending, Tezuka devotes two pages to an extraordinary dream sequence in which Atom, saddened that he has no parents to show his school report card to, bemoans his state on a bridge:

"Wish I had a mom 'n dad like everyone else," he says, staring at a pair of ducks and their young. He then envisions the human faces of his mother and father in space. But the fantasy begins to unravel when images of gears and a robotic mouth appear. When he leaps at a giant robot that looks like a kneeling woman, it crumbles at his touch as he cries out "Mom!!" The sequence ends with Atom prostrate in the rubble, crying for a mother.

Such images would have been very familiar to the young *Tetsuwan Atomu* readers who survived the destruction of World War II. Though Tezuka was not orphaned himself, he identified deeply with Atom, with his spiky head and philosophical nature. In an essay accompanying his collected works published in 1969, Tezuka wrote: "I love Atom more than a son. That's because as

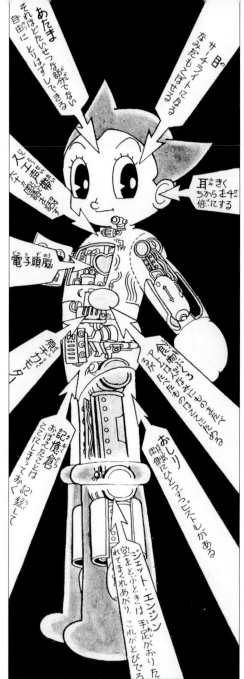

© Tezuka Productions

This illustration by Tezuka shows Atom's inner mechanisms—including his atomic-powered motor—and lists his various powers. Atom had 100,000 horsepower strength and could fly at speeds of Mach 5 with the rocket jets in his arms and legs, shoot bullets with machine guns in his backside, see in the dark with his searchlight eyes, hear things miles away with ears 1,000 times more sensitive than humans, understand sixty languages, and see into the hearts of people and discern whether they are good or bad.

my other self for twenty-odd years from the end of the war, he went through the same experiences and development as I did. When I see Atom, all kinds of memories and experiences from half a lifetime come back to me."

The Once and Future Ambassador

Tezuka was a pioneer in many ways, developing epic-length comics that ran to thousands of pages; he also created a new genre of comics for girls. A workoholic, he drew an estimated 150,000 manga pages during his career, and even turned down an invitation from Stanley Kubrick to be the production designer on *2001: A Space Odyssey* because he was too busy. He also inspired a generation of manga artists who went on to become top names in the industry, such as Fujiko F. Fujio, known for the hit *Doraemon* series about a nearly omnipotent robot cat devoted to helping a pathetic schoolboy.

"Tezuka's influence is endless," says Toshio Naiki, a grizzled Tokyo real-estate agent in his late sixties who houses the largest comic collection in Japan at over 180,000 volumes in three floors above his office. "Many of the early postwar manga had similar themes of space, robots and scientists as well as story structure, and it was obvious they were influenced by Tezuka." He adds, "To put it nicely, I'd say they were influenced. Not so nicely, I'd say they copied him." When the great storyteller died in 1989, the *Asahi Shimbun* newspaper eulogized him thus: "Without Dr. Tezuka, the postwar explosion in comics in Japan would have been inconceivable."

Atom was almost as influential as his creator, spawning a universe of Japanese stories in manga and anime that centered on robots. His adventures certainly paved the way for the genre of giant robots to become a global modern staple of children's fantasy.

He may not be the most popular anime character in Japan these days (ranking twenty-third in a 2005 nationwide poll by a TV network) or even the most popular of Tezuka's myriad characters (it is the rogue surgeon Black Jack, according to Junichi Murakami, curator of the Osamu Tezuka Manga Museum). But the child robot has an unmatched endurance. Besides the manga and books, there have been three TV series (the most recent in 2003), video compilations, feature films and countless items of

merchandise and product endorsement to ensure Atom will be loved and remembered for years to come.

And he had an incalculable effect on robotics in Japan. Just as manga artists readily acknowledge their artistic father in Tezuka, numerous Japanese roboticists trace their first childhood interest in robots to Atom. "Atom affected many, many people," says Minoru Asada, a leading Japanese roboticist at Osaka University who was captivated by Tezuka's grand philosophical themes as a child. "I read the cartoons and watched the TV program.

I became curious to know what human beings are. I still am . . . and that's why I build robots."

Nagoya University robotics professor Toshio Fukuda, who designs "brachiator" robots that ape the motion of monkeys swinging through branches, agrees. In his book *Testuwan Atomu no Robotto Gaku* (Tetsuwan Atom Study of the Robots), he writes: "Although the hurdle of his seven powers is a little too high, the desire to create a robot like Atom exists among Japanese roboticists in varying degrees." He points out that Honda's bipedal humanoid robot Asimo is the direct result of this desire.

One of Atom's greatest contributions to the development—and commercialization—of robots in Japan is the fact that he serves as an almost universal reference point for people inside and outside of robot labs. Atom is a shared ideal, a medium through which scientists and the public can communicate. "Say the word 'robot,'" writes roboticist Fukuda, "and the image of Atom will pop into the heads of most Japanese."

One of the many examples of this is a full-page newspaper ad by Mitsubishi Heavy Industries that features Atom, his sister Uran and Dr.

Atom's popularity over the years translated into a lasting merchandising phenomenon in Japan, as seen in a rash of goods produced to coincide with the date of his creation (April 7, 2003) in the comic series. Here, vintage 1960s Atom toys from Japan's largest collection, owned by Tokyo resident Yutaka Izawa, range from a soft vinyl Atom doll (right) to a tin piggy bank featuring Atom and his robot sister, Uran.

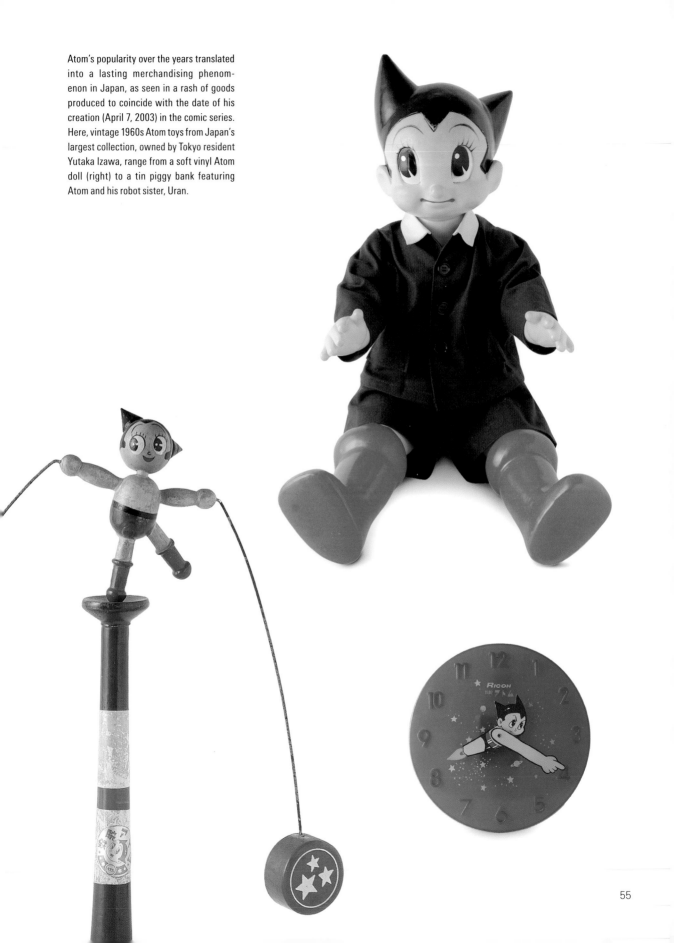

Ochanomizu introducing the household robot Wakamaru. The copy muses: "Why have robots advanced so far in Japan, and why are they so loved? Robot researchers reply unanimously: 'Because of Atom.'"

Science fiction writer Hideaki Sena believes Atom's role as intermediary goes even farther, spanning the gap between fantasy and reality. "We may be able to gain a realistic view of the environment for robots in Japan by thinking of robot stories as interfaces between culture and science," he writes. "Images are being passed back and forth between fiction and real-life science, and these two realms are closely interconnected. This is perhaps the legacy of Astro Boy."

Robotics researchers outside Japan have acknowledged Atom's significance. In 2004, he was inducted into the Robot Hall of Fame at Carnegie Mellon University's School of Computer Science as one of only nine real and fictional robots so honored. The jury, which included MIT's Rodney Brooks and Arthur C. Clarke, decided Atom's claim to fame was being "the first robot with a soul." If that is the key to his popularity, Atom's star is showing no signs of dimming. He continues to rocket through the imagination of popular culture and high technology, leading the way to a better future where robots coexist with humans. More than half a century since he sprang from Tezuka's boundless imagination, Atom is still an ambassador.

CHAPTER 4 Seven-story Samurai

What if one day you suddenly had superhuman power.
How would you use it?

—Koji Kabuto, *Mazinger Z*

Taro is an average schoolboy living in an average suburb. Like a million other kids, he leads an uneventful existence of schoolwork and household chores. Until the day, that is, that mechanical space monsters land in Tokyo.

In the service of an alien empire, they embark on a campaign of destruction to conquer the planet. Taro's father, an eminent scientist, is fatally wounded in the onslaught, but before dying hands him documents relating to a top-secret project. As the capital burns, Taro opens the files. They are blueprints—an operating manual for a giant robot! It is a device of awesome power. It is also humanity's only hope. And Taro is the pilot. . . .

Fans of robot anime and manga will know this plot by heart. It is the formula for innumerable Japanese stories involving "super robots"—titanic anthropomorphic machines of blockbusting might. Often associated with an ancient civilization and a brilliant scientist, the super robot is the ultimate warrior and weapon.

They have names like Brave Raideen, God Sigma and Gaiking. Designed to protect civilization from evil alien forces, they bristle with breathtaking arsenals: gargantuan axes, swords and laser cannons, not to mention fantastic powers like Breast Fire, Teleportation Punch and Antigravity Storm. They sometimes divide into smaller robots or vehicles, and when they do, they are usually piloted by daring Japanese teenagers. They may also transform into cars, jet fighters or spaceships.

Super robots are the apotheosis of the Japanese concept of *meka* (also known as "mecha"), an abbreviation of "mechanism" that denotes machines real or fictional, everything from motorcycle engines to the baroque exoskeletons of battle robots. As they followed in the footsteps of Mighty Atom, the legions of super robots that rapidly proliferated in manga, anime and toy store shelves in the 1970s grew increasingly detailed, realistic and mechanical.

But while Atom was a robot boy who wanted to become human, super robots are essentially tools for human heroes. They have no will of their own and certainly don't share Atom's existential angst. But that hasn't stopped them from becoming an entertainment and merchandising industry worth billions of yen. After all, what boy can resist the idea of skyscraper-sized machines bristling with arms and battling it out like the samurai of old?

Remote-controlled Crimefighter

Japan's first super robot was a towering, needle-nosed humanoid machine called *Tetsujin 28-go*, or *Ironman No. 28* (known as *Gigantor* in some countries). Artist Mitsuteru Yokoyama, who was inspired to become a cartoonist by Osamu Tezuka, began serializing the manga in the iconic boys' magazine *Shonen* in 1956.

Ironman is a clunky steel titan secretly created by the Japanese military in the final days of World War II in a desperate bid to avoid defeat. Twenty-seven times they fail to produce a workable prototype, but the twenty-eighth attempt becomes a three story-tall, remote-controlled machine with a head shaped like a knight's helmet, incredible strength and a jet backpack that rockets it through the air at a neck-snapping Mach 5. Though Ironman doesn't save the Japanese Empire, the twenty-five-ton monster is later put to work clobbering criminal gangs and enemy robots around the world. Ironman is controlled by its inventor's son, whiz kid Shotaro Kaneda, a twelve-year-old detective who wears a blazer, shorts and a pistol.

In contrast to Mighty Atom, Ironman has no artificial intelligence or will of its own, and can be used for good or evil, depending on who is holding the remote. But like its predecessor, Ironman was inspired by wartime experience: the devastation of the U.S.

The first series in the giant robot genre, Mitsuteru Yokoyama's *Ironman No. 28* (right) roared onto the pages of manga magazine *Shonen* in 1956 with its cries of "GWAAA!" Controlled by boy detective Shotaro Kaneda (above, in the latest anime retelling), Ironman is an unstoppable crime-fighting tool.

air raids that Yokoyama witnessed when he was eleven.

"When I was a fifth-grader, the war ended and I returned home from Tottori Prefecture, where I had been evacuated," he wrote in the magazine *Ushio* in 1995. "The city of Kobe had been totally flattened, reduced to ashes. People said it was because of the B-29 bombers, the so-called 'Flying Fortresses' of the sky. As a child, I was astonished by their terrifying, destructive power."

In the first installments, Ironman runs amok like Godzilla, demolishing buildings, army tanks and other robots. The gritty stories feature trenchcoated gangsters, gun battles, dramatic car chases and more plot twists than a mountain road. One of Yokoyama's first jobs required him to watch over two hundred foreign movies, and the influence of cinema is clear in Ironman's breathlessly drawn frames.

The idea of a boy operating a giant robot via a portable controller was immensely appealing. As Yoshihiko Nakamura, a childhood fan who is now a humanoid robot researcher at the University of Tokyo, writes, "The story and fast-paced pictures were like scene changes in a well-edited action movie." Ironman No. 28 was made into a black and white animated TV series in 1963; radio shows, theater plays, films (including a live-action version in 2005) and television series followed. It also inspired generations of imitators and an entire manga and anime genre of giant robots.

The Genie Unbottled

Another of Japan's greatest manga artists took the super robot genre a giant step further in 1972 with the launch of the hugely popular *Majinga Zetto* (*Mazinger Z*, also known as *Tranzor Z*).

Go Nagai was a maverick young cartoonist whose unprecedented use of eroticism, extreme violence and grotesquerie in children's comics made him the enemy of PTAs across Japan. His *Harenchi Gakuen* ("Shameless School") detailed shenanigans in the classroom among perverted teachers and students. Growing up in the remote city of Wajima, Nagai had read both Ironman and Mighty Atom. As a young artist he dreamed of creating his own robot manga, but lacked an original concept—until a revolutionary idea came to him from a very ordinary experience.

While waiting to cross a busy street, Nagai contemplated the backed-up traffic. The drivers must be wishing for some way to get around all those other cars, he mused. Then he had his eureka moment: "A robot that a person can ride in and control like a car!" he thought, imagining one of the cars suddenly sprouting legs and arms and striding through the traffic jam. Instead of simply controlling a robot the size of a building from afar, the hero could pilot it from within its body in a powerful union of man and machine.

Nagai put the concept to use in his *Mazinger Z* manga series in *Shonen Jump*, and a Fuji TV animated series. Both star Koji Kabuto, a rash, orphaned schoolboy who rides a motorcycle. When an apparent earthquake reveals his grandfather's secret underground laboratory, Koji stumbles upon Mazinger Z, a seven-story robot with glowing eyes. Horn-like protrusions on its head and chest suggest both samurai armor and the stag beetles kept as pets by Japanese boys. His grandfather has been mortally wounded, but tells Koji that his creation will make him a superman.

Koji, whose last name means "helmet," pilots the robot by docking a small hovercraft in its head; connected to the controls of Mazinger, he essentially becomes its brain. This dramatic fusion of multiple machines to produce a single giant warrior robot would spawn a host of imitators and is the precursor to the combining and transforming robots that characterized the mecha genre.

Mazinger's name itself comes from *majin*, a demon god or genie, and like Ironman it is a tool that can be used for good or evil: when Koji first tries to drive the robot, it smashes houses and cars indiscriminately. Naturally, the military is called in but their tank shells simply bounce off Mazinger's Chogokin Z metal skin, a fictitious super alloy. Though Koji is briefly tempted by the darker potential in his newfound omnipotence, he resolves to put Mazinger to work fighting evil hordes of robot monsters created by Dr. Hell, a scientist who discovered the basis of Mazinger technology while excavating a lost civilization on the island of Rhodes with Koji's grandfather.

Nagai's most important contribution to the genre was the idea of the robot as a dynamic entity that could join with other machines and exist in symbiosis with a human: by working together as pilot and vehicle they were an unstoppable force. Battling evil invaders every week may have also had a deeper social

Manga maverick Go Nagai had his hero Koji Kabuto drive the giant robot Mazinger Z (far right) instead of control it from afar. The concept proved explosively popular, spawning countless other man-machine combinations as well as a hit series of *Chogokin* die-cast zinc metal alloy figures, reissued by toymaker Bandai Co. in 2002 (above).

resonance. "Practically all the mecha anime plots were SF metaphors for refighting World War II to defend Japan (and Japanese cultural traditions) against the invading armies of Western social influences," writes anime critic Fred Patten.

By 1977, TV was overflowing with imitations carrying titles like *Yusha Raideen* (*Brave Raideen*), *Chodenji Robo Combatora V* (*Combattler V*) and *Wakusei Robo Dangado Esu* (*Dangard Ace*). The latter, a space ship that could turn into a six-hundred-feet tall robot, was created by legendary manga artist Leiji Matsumoto of *Uchu Senkan Yamato* (*Space Battleship Yamato*) fame and was one of the first transforming robotic giants.

As the super robot genre rapidly grew—by the mid-1980s, over forty different giant robot shows had appeared—so did the market for super robot-related toys. Nagai's Chogokin super alloy became synonymous with the die-cast zinc alloy figures produced by toy maker Popy Co. The Mazinger Z toys were solid, powerful-looking and posable. Chogokin robot toys proved so popular that Japanese labs received serious inquiries about whether the alloy existed, according to Nagai. They are still made today by global toy giant Bandai Co.

Several super robot series were made for preexisting toy lines, and some were thinly disguised TV commercials. With their plot lines predicated on a menace-of-the-week where the bad guys always lose, the heavily commercialized super robot genre was ripe for innovation. The man who seized the opportunity created a phenomenon that has become as gargantuan as any super robot.

The phenomenon is known as Gundam.

An Empire of Mobile Suits

To fans of the genre, Yoshiyuki Tomino is a deity in the anime firmament. His 1979 series *Kido Senshi Gandamu* (*Mobile Suit Gundam*), created a sprawling giant robot epic that has inspired generations of Japanese.

Gundam is a remarkably complex space saga with multiple sequels and alternate histories that has been called the *Star Wars* of Japan. The original series, set in "Universal Century" year 0079, features giant robots as weapons in a grand war between human space colonies after overpopulation on Earth sparks famines, epidemics and conflicts

Hero Amuro Ray is a teenage engineer living in an Earth Fed-

eration orbital habitat that comes under attack by giant, faceless Zaku humanoid robots from the Principality of Zeon, a rebel colony loosely modeled on Nazi Germany. In the chaos, he discovers an experimental federation robot that his scientist father has designed, along with the operating manual. Amuro leaps into the machine and makes short work of the attackers. He soon becomes an ace pilot in the inter-colony One Year War as well as a bitter rival of the Zeon commander Char Aznable.

Amuro's warhorse is the signature robot of the series, the RX-78-2 Gundam—an icon of Japanese animation. In appearance, it was a rather typical sixty-feet tall humanoid robot with samurai-like armor. But Gundam differed from super robots in several ways. For one, Tomino labled it a "Mobile Suit" rather than a robot. Another key difference is that Gundam is not invincible. It has no super powers that serve as ultimate weapons. Its arms are futuristic, but conventional: a beam rifle and sabers, a hyper bazooka and cannons. Finally, Gundam cannot transform into physically impossible alternate configurations.

At sixty-four, Tomino is still developing his epic, busily repackaging *Gundam* TV series into movies at Sunrise Inc., the studio he joined in the mid-1970s. He takes his responsibilities as an animator very seriously—he is said to have spent part of his wedding day storyboarding. He also is known for his outspokenness, and has been described, in a Japanese expression, as a man "whose teeth are unclothed." The forthrightness of his views has tended to make him either loved or hated by Japanese anime fans. Yet it is perhaps this part of his character that enabled Gundam to become such a success.

Tomino's nickname is *Minagoroshi*, or "Kill 'em all," for the frequency with which his animated tales end in mass slaughter and the protagonists' demise. Born in Kanagawa Prefecture during World War II, as a young man he worked on the *Mighty Atom* TV series at Tezuka's Mushi Production. But his ultimate vision of robot animation was starkly different to Tezuka's. He went on to work on a number of super robot shows in the 1970s, directing *Brave Raideen*, but chafed at the constraints of the formulaic plots. He also became fed up with what he describes as "making advertising films for toy stores."

"I wanted to make a science fiction, futuristic movie in which humanoid robots are used as tools," the director says in the

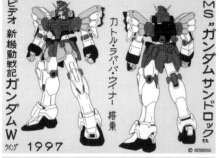

The various Gundam series are characterized by obsessive attention to mecha design detail, as seen in these robots from *Mobile Suit Zeta Gundam* (1985) and *Mobile Suit Gundam Wing: Endless Waltz* (1997).

cramped offices of Sunrise, located in a quiet Tokyo suburb. "I first thought of these humanoid robots as vehicles. I thought they could be giant vehicles the size of fighter aircraft or tanks and be mass produced."

Some have speculated that Tomino's Mobile Suits were inspired by Robert Heinlein's 1959 SF novel *Starship Troopers*, in which powered armor exoskeletons are used by troops in a mobile infantry battling insect-like aliens. But Tomino says this is incorrect. He arrived at the concept via simple reasoning: In the terrifying void of space, it would be comforting for people to work alongside machines with a human shape. He associates this desire for anthropomorphic robots with religious history in Japan, particularly the worship of the carved images of Buddha in countless temples throughout the country.

Spacesuits and humanoid space tools are thus the origins. "Mobile Suits in Gundam are just weapons," adds Tomino. "Of course they can seem more than just a tool because Mobile Suits have a human form."

Instead of making the robots heroes, the series focuses on its human characters and their heroic roles in the space war. But can they become heroes only by encasing themselves in a massive machine, sealed off from the environment? "Think of it like a Formula One driver," Tomino says. "It's a symbol of the relationship between human and the tool. The characters themselves are just common people. But influenced by a machine, there is something heroic about them."

The most striking feature of the Gundam series is its realism. The Mobile Suits obey basic laws of physics, albeit with license for space fantasy. The robots have names with complicated alphanumeric prefixes. They are products of industry, with prototypes and mass production — manufactured commodities that can be repaired or replaced when damaged.

Furthermore, the robots in the Gundam universe are not one of a kind like Ironman No. 28. They come in myriad forms and with multiple functions, and have evolved into unique creatures of mecha design. Some resemble birds with fanciful winglike appendages, others are blockish and heavily armored like tanks. Yet they are basically tools in a military arsenal, industrial war machines created by governments and corporations. They have no romantic origins in some lost civilization like Mazinger Z.

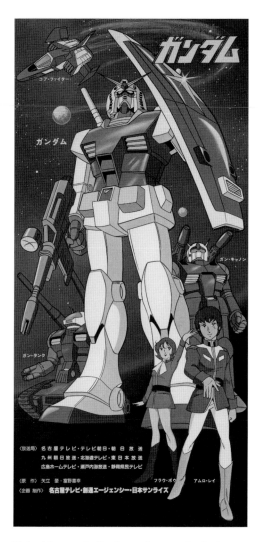

ガンダム

ガンダム

コア・ファイター

ガン・キャノン

ガン・タンク

〈放送局〉名古屋テレビ・テレビ朝日・朝日放送
九州朝日放送・北海道テレビ・東日本放送
広島ホームテレビ・瀬戸内海放送・静岡県民テレビ

〈原 作〉矢立 肇・富野喜幸
〈企画 制作〉名古屋テレビ・創通エージェンシー・日本サンライズ

フラウ・ボウ   アムロ・レイ

Diehard fans made Gundam a Japanese institution despite the fact that the initial series, the now-classic *Mobile Suit Gundam*, had low TV ratings. When their space colony is attacked by Zeon forces, friends Amuro Ray and Fraw Bow become embroiled in a war fought by people in giant machines.

There is plenty for mecha fans to chew on, though: the amount of available technical details on Gundam is mind-boggling. The Gundam Museum outside Tokyo features meticulous treatises about the fictional robots that could pass for scientific papers. Take one display that explains the mechanical challenges in the development of the Gundam RX-78-2's 60mm caseless triple-barreled vulcan cannon, manufactured by TOTO Cunningham Inc. Somewhat surprisingly, the museum's displays are in English, perhaps to impart an authoritativeness implicit in the global language.

Incredibly, given the popularity of the series today, audiences at first responded coolly: low ratings prompted the cancellation of the initial 1979 show. Gundam could have died there, had not a group of diehard fans lobbied for more. Following the release of a three-part movie compilation, the franchise grew over the next two decades to include nine television series, over ten films and many comics, novels and video games.

But one of the most remarkable aspects of the Gundam phenomenon is its merchandising power. There is an empire of related goods, including action figures, games and collectibles. Bandai, Japan's largest toy maker, relies on the Mobile Suits as its major revenue source, raking in ¥42.8 billion in 2004, most of it from Gundam model kits, which are snapped up by Japanese males, many of them zealous hobbyists called *otaku*. Since 1980, over 360 million units of some five-hundred different kits have been sold, making the models the biggest hit in Japanese hobby industry history. The appeal lies in the realism and the variety. "We try to imagine what they'd be like if they actually existed," says Sei Okazaki, a Bandai manager with every boy's dream title: Deputy Chief Gundam Officer.

Tomino never expected his Mobile Suit idea would bloom into such a massive market, and now tries to avoid the Gundam mania. He believes anime is an art form and claims he is unhappy with the commercial industry it has spawned. "The current boom in otaku culture is like an ocean," he says. "And I have to keep my head above water. If I sink into it, I'll die."

Tomino, however, was "very pleased" to see how his creation's popularity was manifest in another, unexpected field: modern art. In an extraordinary 2005 exhibition in Osaka and Tokyo entitled "Gundam: Generating Futures," young Japanese artists presented their impressions and visions of the Gundam universe. Some

Gundam fandom knows no limits. "Because I encountered Gundam in my adolescence, without knowing it it's become part of my flesh and blood," said Tokyo-born artist Hisashi Tenmyouya, whose surrealistic "RX-78-2 Kabuki-mono 2005 Version" reflects his signature mix of traditional motifs with modern imagery. The painting was featured in the exhibition "Gundam: Generating Futures."

works were sketches of battles involving countless Mobile Suits, and others were abstract cartoon characters evoking the RX-78-2 or feverishly detailed renderings of it.

One of the most striking pieces was Hisashi Tenmyouya's *RX-78-2 Kabuki-mono 2005 Version*, which depicted the title robot in a traditional Japanese screen painting—but with a coiled dragon emerging from its cockpit. Tenmyouya's vision came from an urge to fashion his own Gundam. "Because I encountered Gundam in my adolescence, unconsciously it's become part of my flesh and blood," he said in an interview for *Art It* magazine. "Etched in my memory is an episode from my first year at high school when I asked the girl I was going out with at the time, 'Who is your favorite celebrity?' The answer she gave wasn't a celebrity but [Gundam pilot] Amuro."

Gundam popularized the gritty "real robot" genre featuring giant robots that are realistic pieces of hardware, often sporting oversized weaponry, rather than superheroes. And myriad new series featuring real-type robots continue to emerge in a variety of cross-polinating media. Shoji Gato's *Furu Metaru Panikku!* (Full Metal Panic!), which features saucer-eyed schoolgirls and giant combat mecha, started off as a light novel serialized in a magazine and has become an animated TV show and a manga comic. Other titles like Sony Computer Entertainment's *Kenranbutosai*, a mecha adventure story about pirates on a watery Mars, began as video games and have inspired animated TV series. Successful shows, in turn, are often compiled into movies and re-released. In the ultra-competitive video game market, one popular combat mecha title is Armored Core, a hit multiplatform series in which players pilot customized mecha "cores" in a hardboiled, post-apocalyptic world.

In a cinematic sequence from Armored Core's latest installment, "Last Raven," that is reminiscent of *The Terminator*'s future dystopia, the cores are portrayed as almighty military machinery that tower over human soldiers in a shattered cityscape. (This is perhaps the extreme end of the real robot spectrum, where anthropomorphism submits to the aesthetic of the gun.) The robot has become little more than a lethal collection of machine parts. Indeed, "robot" seems a misnomer, and many Japanese SF worlds describe their humanoid mecha with invented names like

Gundam creator Yoshiyuki Tomino's elaborate robot saga became a merchandising bonanza. Gundam products, such as Bandai's 1/60 Strike Freedom Gundam (Lightning Edition) from *Mobile Suit Gundam Seed Destiny* (right), generate some ¥42 billion in annual sales.

cores. Though they are clearly the direct descendants of Tomino's RX-78-2 robot, and by extension the pixie-like Mighty Atom, one wonders what the little robot and his peace-loving creator Osamu Tezuka might say.

Ghosts in the Machine

If Gundam represents the maturation of the super robot genre, then *Shin Seiki Evangerion* (*Neon Genesis Evangelion*), one of the most popular giant robot anime of the past decade, is its midlife crisis. The groundbreaking series was created by writer-director Hideaki Anno and a group of fellow super otaku who made films together in college before setting up Studio Gainax in 1984.

Born in Yamaguchi Prefecture, Anno loved mecha stuff as a kid. He would take apart discarded TV sets and pore over their vacuum tubes, sometimes creating fanciful assemblages out of junk electronics. He also loved enormous vehicles like the warships he read about in boys' magazines or the spaceships in Matsumoto's Space Battleship Yamato. Anno was also influenced by manga such as Mazinger Z, while Mobile Suit Gundam and Tomino's later *Densetsu Kyojin Ideon* (*Legendary Giant God Ideon*) were direct precursors to *Evangelion*.

But the series represents a departure from the dominant real robot genre. The formula of boy-meets-robot-and-fights-evil remains, but *Evangelion* elevates it into a superbly animated drama of Wagnerian scale.

It was first broadcast only months after the Aum Shinri-kyo doomsday cult's sarin gas attack on the Tokyo subway in 1995, which shattered Japan's image of itself as a relatively crime-free society. Another background factor was Anno's own depression. Fraught with unrelenting anxiety and interwoven with existential questions—all to the highbrow accompaniment of Beethoven and Bach—the show became a massive hit and inspired an army of devotees.

The story begins in 2015 after a cataclysm eradicates half of humanity and precipitates global conflict over resources. When the new city of Tokyo 3 comes under attack by mysterious giant bio-mechanical monsters and other threats called Angels, a fourteen-

Though inspired by Gundam, Hideaki Anno's *Neon Genesis Evangelion* was strikingly original in its detailed psychological analysis of characters like troubled protagonist Shinji Ikari (right, front) and robots that look more like animals than machines.

year-old named Shinji Ikari is pressed into service as a pilot for U.N. special agency NERV (led, of course, by Shinji's scientist father).

The only thing that can stand up to the Angels is Shinji's vehicle—the Eva Unit 01. It is an enormous anthropomorphic living weapon, more animal than machine. "The term 'robot' originally referred to the idea of a man created by artificial means," NERV scientist Ritsuko Akagi tells a reluctant Shinji in a manga version. "This indeed is a synthetic android . . . an adaptable human construct and our decisive weapon. Its code name is Evangelion." The Eva 01, with its narwhal horn and monstrous jaw reminiscent of Mazinger Z's baleful expression, was conceived after mecha designer Ikuto Yamashita was told to imagine a demon. The design theme, according to Yamashita, was "enormous power restrained."

Despite their armor plating and trailing umbilical power cables, Eva 01 and the other Evas created by NERV are more than mere mechanisms. Half-biological, the Evas contain souls of the child pilots' dead mothers with which they synchronize mentally after being immersed in an amniotic-like fluid in the cockpit. (Once again, the psychic plight of orphans is tied to robots.)

For its fans, *Evangelion* is far more than an SF adventure. The giant humanoid biomechanisms are secondary to the convoluted emotional struggles of Shinji, Rei and the other children. Detailed character introspection and psychoanalytic passages alternate with episodes of graphic violence, making this more of an anime for adults than kids.

"I belonged to the generation that grew up hearing that science is wonderful, science is great, science rules the world," Anno told the Japanese edition of *Popular Science*. "We also heard that due to the benefits of science we'd definitely have a bright future. But then pollution became a social problem. We also had the oil shock, and then people started saying totally different things, like the future didn't look so good. We wondered whether our future with science would be good or bad."

Although its characters and mecha start off with a psychological and technical realism inherited from Gundam, Evangelion adds an obsessive mysticism. The series is laden with Judeo-Christian references and symbolism, from its Greek title, meaning "gospel," to the crucifixion of the Angel Lilith with the legendary Spear of Longinus, which pierced Jesus on the cross.

Some might say that Evangelion is deliberately incoherent.

If so, that echoes the schizophrenic malaise at the heart of the story. Interviewed in 1996, character designer Yoshiyuki Sadamoto admitted: "Our aim was to be the antithesis of all the giant robot animated shows around us." Artist Takashi Murakami sees the series as representative of its country of origin, writing in the catalog to the seminal exhibition *Little Boy*: "Shinji's identity crisis, apparently a reflection of director Anno's own psychological dilemmas, epitomized the difficult obstacles faced by postwar Japan, a nation that recovered from the trauma of war only to find itself incapable of creating its own future: like Shinji, Japan is probing the root cause of its existential paralysis."

Science Imitating Art

Born of war, giant robots are art imitating life, the ascendancy of the machine. But makers of actual robots also imitate art. The body designer for the final version of the HRP-2 Promet humanoid (see page 153), developed by Japan's National Institute of Advanced Industrial Science and Technology along with about twenty universities and major robot companies, was Yutaka Izubuchi, an anime mecha designer of Gundam fame.

The influence goes beyond looks. Sakakibara Kikai's Land Walker, the eleven-foot legs and cockpit with no practical value whatsoever, is an early realization of "real robot" machines that sprang from Gundam. And, in introducing the drivable rescue robot Enryu T-52 Advance, a ten-ton juggernaut on treads that can pick up cars with its two arms while operating in disaster zones, developers Yoichi Takamoto and Yasuyoshi Yokokoji write: "Of course it doesn't fly like Ironman No. 28, but somehow resembles his powerfulness when he's rescuing people."

"What Japan can truly be proud of is monozukuri and technology," says *Evangelion*'s Anno. Robots, he believes, can represent Japan. In fact, he thinks Honda's P2 bipedal humanoid should have lit the torch at the 1998 Nagano winter Olympics.

Decades after mighty mechanical warriors first appeared in the pages of comic books and on screen, their hold on the Japanese conception of robots remains unbroken. Doubtless ever-more amazing machines will continue to leap from the Japanese imagination—extreme robots in size, realism and power.

Sakakibara Kikai's towering Land Walker is a realization of the drivable giant robots of manga and anime. The shuffling, two-legged metal cockpit can fire soft balls from a cannon, runs on a 250 cc engine and is controlled via four pedals.

CHAPTER 5 Of Walkers and Workers

*The human motor system—arms and legs—
is a type of servomechanism.*

—Ichiro Kato

Over the postwar years, as some minds—like those of Tezuka and Tomino—took imaginary flight to a time of robot children and battling robot behemoths, others bent to the more difficult task of actually designing and building humanoids. At Tokyo's Waseda University, research was taking place that would result in the world's first full-scale humanoid robot, introduced in 1973. The credit would go to a scientist nicknamed "Professor Ochanomizu," after the genius inventor who is Mighty Atom's foster father in Tezuka's stories.

Ichiro Kato, a specialist in control engineering and one of Japan's most famous roboticists, began building humanlike machines in the 1960s. The soft-spoken man was a visionary with an omnivorous appetite for knowledge, equally passionate about philosophy and physiology as engineering, who persevered with his research even when other academics dismissed it as child's play.

Kato viewed the human body in mechanistic terms; he believed human abilities could be furthered with the application of machines. He studied factory automation and servomechanisms, which are important robotics devices that permit automatic control. Duplicating the human hand seemed to be a logical point of departure for research.

His 1964 Waseda Hand was driven by a single motor that used pressure sensors on the fingers to automatically choose between pinching or gripping objects. The mechanism was relatively crude

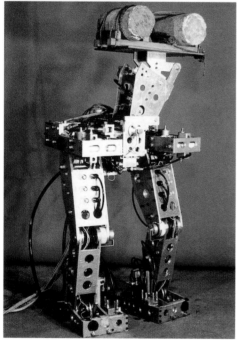

Early humanoids at Tokyo's Waseda University. Completed in 1972, the WAM-4 torso (top) had visual and tactile sensors and could transfer objects from one hand to the other. Loaded with heavy weights, the WL-5 walker (below) was controlled by a minicomputer and took forty-five seconds per step, but achieved static walking.

at first but the hand evolved rapidly over the years. The goal of Kato and his students was to produce a working artificial hand for amputees. Electromyography, which measures the electrical activity in muscles, was employed so that nerve signals from a human arm could be converted to gripping or pinching actions.

By 1968, the WH-4P clinical model could also provide thumb sensor pressure feedback to an amputee via electrical pulses in the skin. The WIME Hand commercial model, launched in 1978 and manufactured by prosthetics maker Imasen Engineering, added pressure feedback from three fingers, and came with simulated skin over a steel armature. It was sold to amputees in Japan until 2004.

WABOT: From Prosthesis to Person

Kato was also profoundly interested in locomotion. He spent hours studying slow-motion films of animal motion, but his main interests were human walking and the structure of the human body. His ultimate goal, in fact, was not only getting a robot to move like a person but recreating man in robot form.

Until then, robots had been relatively simple bundles of sensors and motors that rolled on wheels, like the electronic tortoises with radio tubes for brains that were developed around 1950 by British psychologist W. Grey Walter. Twenty years later, the more sophisticated Cart and Shakey robots built at Stanford University and Stanford Research Institute were controlled by room-sized computers and used primitive computer vision to navigate their environments on tires—very slowly. But no one had really tried to build a robot in human form.

Having found funding through his strong connections to Japanese manufacturers and industrial robot makers like Yaskawa Electric, Kato applied his prosthetics know-how to build "machines to perform manual labor in lieu of persons," as he wrote in a 1985 history of his research. He first developed robot limbs, then set about joining them together. He began with arms, which grew from simple appendages with rubber muscles to ones with computerized control and automatic object grasping. Two such arms, joined to a torso, were called the WAM 4. Next was the WL-5, a pair of minicomputer-controlled robot legs topped by swinging weights that altered its center of gravity. It took a full forty-five

Ichiro Kato's WABOT-1, developed in 1973, was the world's first full-scale anthropomorphic robot. It could speak Japanese, learn about its environment through visual and audio sensors and carry objects in its hands. A fusion of the WAM-4 and WL-5, WABOT was said to have the mental abilities of a child of one and a half.

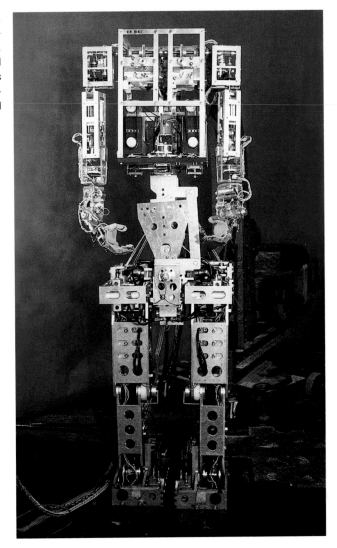

seconds to complete each step, shifting its weight from one foot to another in what is known as static walking.

Together, these limbs became WABOT-1, an anthropomorphic robot introduced in 1973 with integrated vision, motion and conversation systems that was the first of its kind in the world. It lacked a head and its body looked like a collection of computer console innards, but it had artificial eyes and ears and was said to have the intelligence of a human child of one and a half.

Kato and other researchers, however, already had their eyes on a long-term goal: a personal robot called My Robot that could work in the service industry and handle information as ably as humans. A helpful artificial man, it would be the third stage after industrial robots and computers in what Kato saw as an evolutionary

WABOT-2, seen here in 1984, was designed to take on the intelligence and dexterity challenge of reading and playing music. It could scan scores with its camera eye, play them on the organ and engage in conversation. A modified version called Wasubot played Bach with an orchestra at Japan's Expo '85.

Ichiro Kato

process. Though his sluggish WABOT had zero practical value, it fired the imaginations of many.

"At that time, a robot was like a fantasy," says Shuji Hashimoto, co-developer of the WABOT vision system and now director of the Waseda University Humanoid Robotics Institute, which boasts eight professors and dozens of researchers. "Of course we knew what an industrial robot was, but only that they were used in factory automation. In my childhood, I read some manga about [humanoid] robots but I didn't realize that it had become time to make them." He remembers asking Kato and his team: "What will you make, Mighty Atom or Ironman No. 28?"

The Well-Tempered Clavier

Kato asked other research groups to help improve his automaton. He recruited Waseda specialists from fields like information processing, AI, and speech recognition. WABOT-2 was developed in 1984, eleven years after its predecessor, and it was intended to show that robots could perform intelligent and somewhat practical tasks. Equipped with a single large video camera for a head, it had limited AI but could read sheet music, play the organ and accompany a singer. It still looked very mechanical, however, with its body of metal cylinders and drooping cables.

To show the world how well Japanese had accepted machines and robots into their lives, a slightly modified version of this electric pianist named Wasubot, with a repertoire of sixteen pieces, read and performed J.S. Bach's "Air on the G String" with the NHK Symphony Orchestra at Expo '85 in Tsukuba, northeast of Tokyo. Emperor Akihito, then crown prince, was in the audience along with thousands of others.

In *Inside the Robot Kingdom: Japan, Mechatronics, and the Coming Robotopia*, author Frederik L. Schodt describes the scene thus: "Wasubot was a humanoid, and when he played it was not a recording, but a provocative performance. Some listeners saw an enormously complex and expensive machine usurping a most emotive human activity, and felt a chill. Others saw a robot—a mechanical mirror of themselves—and heard music that moved them to tears."

"I felt like, 'We did it!'" recalls Waseda's Hashimoto, who collaborated on Wasubot's vision. "It was a very big project for us because so many different teams joined in. Professor Kato's group

Waseda's humanoid robots continue to evolve in the WABIAN series. The first generation (far left) was created in 1995 and had thirty-five degrees of freedom. Atsuo Takanishi's lab is developing WABIAN 2 (left) as an exact model of human locomotion and test bed for medical and welfare equipment to help the aging Japanese population stay mobile.

Takanishi and robotics company Tmsuk joined up to invent a mobile robot that can carry a person and climb stairs, a world first. The WL-16RII is designed to help wheelchair users, and incorporates telescopic poles that move at a leisurely rate of 0.96-1.92 seconds per step. Takanishi calls it a "two-legged walking chair."

made the mechanical control system, we made the vision system, someone else had to make a speech system. Five or six professors were born in that project."

Kato's aim of producing a practical humanlike robot was never achieved in his lifetime; he died in 1994. His dexterous, musical robot was impressive but not useful, and his goal of producing an anthropomorphic machine that could carry out intelligent tasks instead of humans in the service sector remained decades away. Wasubot was basically an entertainment robot in the tradition of Gakutensoku and karakuri clockwork dolls.

Waseda is continuing its four decades of humanoid robot research. The latest product of the university's robot labs is one of the most advanced walkers in Japan, the super-flexible WABIAN-2, which has forty-one degrees of freedom. With its interior mechanisms partly visible, the blank-faced, five-foot tall biped was engineered to walk fluidly with bobbing hips and pivoting ankles that perform unique "S-turns" to avoid obstacles. It is designed both to work in human environments and as a "human motion simulator." Researchers at Atsuo Takanishi's lab believe that as Japan's population ages, there will be greater need for rehabilitation tools, and a humanoid that can mimic human walking will be useful in developing them.

"In what I call Robotic Human Science, humanoid robots are used for scientific research about humans," Takanishi has said. "If

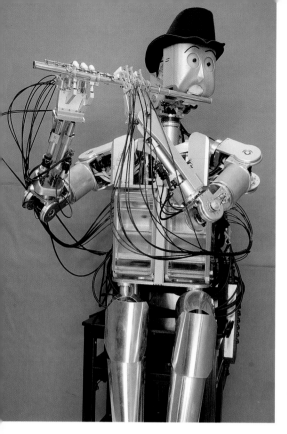

This flutist from the Takanishi lab continues the tradition of Kato's musician robot as well as karakuri automata performers. The WF-4RII simulates human lungs, lips and fingers as well as tonguing and vibrato techniques. It is an attempt to understand flute-playing through robotics.

this advances, humanoid robot dummies could be used instead of humans to test and evaluate products in many industrial fields like medical equipment and therapies, automobiles and clothing."

Such endeavors illustrate Kato's most important legacy—he helped show that building a humanoid robot capable of operating in the human environment was not only feasible but worthwhile. Perhaps not coincidentally, the year after Wasubot's steel fingers brought the music of Bach to life, Honda Motor began research into what would become the world's most sophisticated humanoid robot today, Asimo.

And, meanwhile, Japan's factories were taking part in a quiet revolution that would not only contribute to the country's economic miracle, but would open the doors for robots into many walks of life.

The Robot Kingdom Workforce

Two combatants attack and parry, their red and blue lightsabers dancing in the air. The accompanying soundtrack is the *Star Wars* theme, but this is no movie. The duelists are a pair of arc welding robots, and this elegant swordplay, a modern karakuri spectacle, is how industrial robot maker Yaskawa Electric introduces its "Motoman" series lineup at the world's largest robotics trade fair in Tokyo. Not far away, a realistic robot tyrannosaurus is stomping around a stage in front of a spellbound audience. But the businessmen watching these two mechatronic Jedi are equally entranced.

To the outsider, the industrial robots displayed by some one hundred and fifty companies at the sprawling International Robot Exhibition in Tokyo, which draws one hundred thousand people over four days, might seem like a strange machine ecology. A giant Kawasaki Heavy Industries MX500 heavy-duty manipulator, a metal Godzilla, effortlessly hoists a lime-green motorcycle off the ground and twirls it around. Nearby, a Nachi-Fujikoshi SJ120C LCD panel-handling robot telescopes across the floor and up two stories into the air like an unfolding giraffe. Meanwhile, a yellow Fanuc spot welding arm, bulky but nimble as a monkey, reaches around an auto body part at blinding speed and suddenly halts at a precise point in space over a joint.

These articulated, whirring creatures of steel are making a rare public appearance. They more often toil away unseen in factories

Welcome to the Robot Kingdom: Accompanied by the theme from *Star Wars*, two Yaskawa Electric industrial robots engage in a light-saber duel at a trade fair in Tokyo. Even in the hard world of Japanese factory automation, room is made for fantasy and fun.

across the Japanese archipelago, some with virtually no human supervision, mounting microchips, painting parts and making everything from luxury cars to toilet bowls—and, of course, other robots. They are vital laborers in the world's second-largest economy. Boasting half the world's industrial robot population, Japan is the most automated society on earth, a status it has retained for decades. In fact, Japanese sometimes call their country the "Robot Kingdom" because of these worker drones. Where would be a more appropriate setting for the birth of the world's most advanced humanoid robots?

An American Idea, Big in Japan

A chance meeting between two Americans at a 1956 cocktail party in Connecticut was an important event in the rise of Japan's robot empire. Self-taught inventor George C. Devol Jr. struck up a conversation with Joseph F. Engelberger, a physicist, engineer and Isaac Asimov fan who became fascinated with Devol's patent filing for the world's first programmable industrial robot, a "programmed article transfer" machine. Engelberger went on to found Unimation, which pioneered the industrial robotics field.

Unimation was based on Devol's notion of "universal automation" and his goal of flexible automation in the manufacturing process. The clunky but reliable Unimate, which looked more like a giant gun than an artificial man, could perform a variety of tasks such as manipulating objects or welding. But Unimation's sales team met with resistance among U.S. manufacturers who were slow to recognize the benefits of having robots on the shop floor. The response in resource-poor, export-dependent Japan, however, was very different.

Japan was to implement industrial robotization on a massive scale, quickly learning from, then surpassing the United States in robot development. Author Schodt notes, "No other nation has put so many robots to work in so many industries as Japan, or with so much success."

In 1968, Kawasaki Heavy Industries (then Kawasaki Aircraft) tied up with Unimation to produce the Kawasaki-Unimate 2000, Japan's first domestically created industrial robot. The same year also saw Ishikawajima-Harima Heavy Industries and Yaskawa Electric start selling robot systems for manufacturers. Other Japanese companies licensed robotics technology from the West and competition ensued. By 1978, Japan was producing ten thousand robot systems a year.

The growth was fueled by Japan's post-war export boom, much of it originating in the automobile industry. Robots, with their strength and tireless ability to repeat the same motion over and over, were perfectly suited to automobile production lines. After seeing what the machines could do for automakers, other industries started to embrace robots to make up for skilled labor shortages amid the booming economy.

The 1960s were a decade of invention for robotics, the 1970s of development and the 1980s of growth. In Japanese industry, 1980 is referred to as "Year Zero" for robot automation because that was when the industry took off, growing from ¥78 billion to ¥600 billion by 1991. Nearly everything that could be robotized was. New corporations emerged, including some that would become titans in the global robotics industry.

What's the most obvious upside of the having built the Robot Kingdom? "The automotive industry, for instance, has enjoyed the merits of automation," says Yoshiaki Nakatsuchi of Kawasaki Heavy Industries. "This has led to lower prices for cars. Manufacturers, consumers and robot equipment makers have all benefited."

Without the Robot Kingdom, Japan would not have become the export powerhouse that it is today. Nor would it have been possible for large Japanese companies to produce headline-grabbing, sophisticated and dexterous bipedal humanoid robots like Sony's Qrio—which itself incorporates actuators based on components in its consumer electronics manufacturing robots.

Nonhuman Coworkers

In hindsight, it is not difficult to pinpoint why Japan was able to embrace industrial robots so easily and achieve world supremacy in the field so quickly. National goals like the drive to develop electronics exports, skilled labor shortages and a greater willingness to take investment risks with cutting-edge technology were factors. Another reason was the difference in labor union structure between Japan and the West. Japanese labor groups tend to be enterprise unions based at individual companies and not industry-wide organizations based on a trade or specialty.

"In the U.S. and Europe, if painting or welding workers became redundant they'd have to find a new job elsewhere, but in Japan workers could be transferred to a new job within the company," says Shigeaki Yanai of the Japan Robot Association, an industry body. Losing one's position to a robot was easier to swallow if it didn't mean losing one's job.

The view of Western workers as robot-averse is, of course, a stereotype, despite negative cultural portrayals of robots from *R.U.R.* to *The Terminator*. Michael J. Bomya, executive vice president at robot-maker Nachi's U.S. unit, recalls installing a vision-guided welding system at a GM plant in North America decades ago and having concerns about displacing staff. "When I got it up and running, one of the two workers who had been doing manual arc welding came up to me, pulled a flask of whisky from under his jacket and said, 'Let's drink a toast,'" Bomya remembers. "He said it was the most terrible, dirty, backbreaking job in the world. He was happy that robots had been put on it."

Yet there is a stark difference between how Western and Japanese workers treat their nonhuman colleagues. One memorable observation was made by Unimation's Engelberger after a visit to a Mazda Motor plant years ago: "It's a matter of dishonor if the robot stops on their shift," he said. "They were there on their coffee break, polishing the things. Every machine was shiny, and they

Drones like these Motoman auto parts arc welding robots have helped resource-starved Japan become an export power-house through flexible automation. Manufacturers have reaped huge productivity rewards thanks to robots, turning out better products at lower cost and responding to market changes faster.

knew it was theirs, and they were proud to be able to go home and say they knew how to program it. Now in the States, you run it till it breaks. If you need time off, maybe you kick it till it breaks. The whole attitude is different."

Part of this may be the Japanese work ethic. The Robot Association's Yanai, however, once again points to the influence of Japanese comic books and animated films as a critical difference. "We have an affinity toward robots," he says, "stemming from anime and manga. Robots were our heroes, and they still are. There can be bad robots too, but other robots will always destroy them. That's how it is in *Mighty Atom* or *Ironman No. 28* and the others. From our childhood we think robots are crime avengers."

Yoshikazu Suematsu, the karakuri scholar and dean of the Toyota National College of Technology, traces the roots of this affinity for industrial robots deeper in the past. He believes that centuries of enjoying karakuri dolls performing in settings like religious festivals is what allowed Japan to deploy industrial robots so widely.

Instead of rejecting or fearing assembly robots, then, factory employees would welcome them in ceremonies, even naming them after popular celebrities. In Japanese plants, especially smaller ones, new robots were welcomed with Shinto ritual purification rites, complete with a visiting Shinto priest, and accepted as part of the staff.

Fewer People, More Machines

While other countries are slowly catching up to Japan in the industrial robot race, the Robot Kingdom is still growing—and looking ahead to the next stage in the evolution of the nonhuman factory worker to meet its own needs. As the rapidly aging population dwindles and baby boomers retire, manufacturers are facing massive labor shortages, a greater threat than competition based on cheap labor overseas.

In the automotive industry, robots have not been widely used in tasks that are more complex than joining metal parts together and spray-painting them. But that is changing. Toyota Motor, which uses thousands of robots at its twelve domestic plants and also makes humanoid Partner Robots that play musical instruments, is introducing next-generation two-armed robots that can do more than merely weld and paint; they will take on the complex finishing jobs still performed by humans, like installing seats

Although heads were put on these new Yaskawa Motoman-DA20 robots as a playful gesture for a demonstration, they signal a more anthropomorphic industrial automaton and an engineering solution to the shrinking workforce. Each six-jointed arm can carry out a separate task such as bolt-tightening. Yaskawa has also developed a bartender version that sports a bowtie and mixes cocktails.

and interior fixtures. Nissan Motor, meanwhile, is using laser-equipped automated equipment to inspect all its Fuga luxury sedans instead of a random selection checked by workers.

Aided by computer advances, factory robots are also getting smarter. In an assembly cell at its headquarters plant near Mt. Fuji, Fanuc's new 3D laser vision-equipped Intelligent Robots manufacture other robots—twenty-four hours a day, at a rate of two hundred a month—and autonomously make motion decisions without prior programming. With integrated force sensors, the Intelligent Robots can be used in money-saving applications like bin picking, a long sought-after industrial robot function in which the machine autonomously recognizes and handles parts that have been randomly placed in a bin instead of working with neatly arranged stacks and prior knowledge of the location of parts.

"People have had to do work for robots until now, which is the opposite of the way it should be," says Ryo Nihei, a member of Fanuc's board of directors. "These robots eliminate unpleasant work."

And the latest entry in the Yaskawa Motoman series suggests that industrial robots are not only taking over more and more intricate tasks performed by people, they are adopting their looks as well. The Motoman-DA20 has two arms, each with six joints, specially designed miniature servomotors and is about the size of a man's torso, meaning that unlike larger machines it can work alongside people. Unlike humans, though, it can tighten bolts with one limb while toting parts with the other.

In a demonstration at the International Robot Exhibition, DA20s were dolled up in neck scarves and even installed with mock "heads," giving them a truly humanoid appearance. But across the hall at a prototype robot show, Yaskawa's SmartPal illustrated the potential of highly refined automation technology outside the factory.

SmartPal is an interactive service robot that can autonomously perform practical tasks like retrieving and serving sandwiches on a tray. Moving about on its wheeled base and using its sensors to avoid obstacles, it looks like a floating steel dress. "I was born to work in factories and offices in the future," SmartPal says in its metallic, synthesized voice. "With seven joints, my arms can move just like a human's."

Unlike most industrial robots, its supple limbs incorporate compact control units. Yaskawa says SmartPal could be used for simple assembly, or, since it has voice recognition abilities, as an office receptionist.

Though it can also serve drinks, SmartPal's green touch-panel LCD screen, which serves as a kind of face, would make it less than the ideal bartender, someone to confide with all your troubles. Yet it is the most vivid example of evolution in the machine ecology of industrial robots, a whole new stage in which the human form becomes increasingly adopted for work that requires far more in terms of physical abilities, mobility and intelligence than simple welding or painting. And if next-generation factory robots do begin colonizing service industries in the future, they will find a host of consumer-oriented machines there waiting for them.

Bartender SmartPal's LCD face glows as it waits for an order. Incorporating advanced automation technology, it can serve up food and drinks using its seven-jointed arms. Yaskawa sees such service robots performing routine jobs and supporting people in their everyday lives.

CHAPTER 6 Humanoids at Home

Robots will become the Ford Model T of the 21st century.
—Hirohisa Hirukawa, Humanoid Robotics Group, National
Institute of Advanced Industrial Science and Technology

The robot revolution that brought autonomous machines out of the factory and into people's homes began at the close of the twentieth century with a whining, whirring contraption full of *kawaii* cuteness, Aibo. Choosing a name for a robot dog that meant "buddy" in Japanese (as well as referring to "AI" and "robot") was a natural move by its maker, electronics colossus Sony Co. When the first three thousand Aibo ERS-110 hounds were let off the leash in May 1999 in Japan, they sold out in a mere twenty minutes. (It took four days for the first two thousand Aibos to sell out in the United States.)

Although Sony discontinued making the robots when it folded its entertainment robot business in 2006, the company promised to continue maintenance for a period of seven years. It was, most likely, a wise corporate decision, considering the widespread popularity of the robot dogs. After all, Aibo was the first mass-marketed autonomous robot—combining cuteness with the fun of a pet and the coolness factor of playing fetch with a machine. Loaded with sensors and actuators, Aibo walks, talks, barks, dances, plays music and expresses its six "emotions" on its LED muzzle and by wagging its silicone tail; when ignored, the robo-pup sulks. The final version, the Aibo ERS-7M3, can also wirelessly email its owner digital photos that it takes with its onboard camera. And it can add its own simple comments about people it photographs, thanks to face-recognition technology.

It is hard *not* to find the gleaming puppy endearing as it gambols about on a desk or chomps down on its Aibone—a special color-coded plaything. As it got smarter and more sophisticated with each generation, Aibo seemed more and more like a living being. Though mass-produced, it is customizable, with different movements, sounds and games that could be installed at will. (Owners can also install Aibo with software that gave it a Kansai accent—the distinct Japanese dialect of the region around Osaka.) As its personality develops under the care of its owner, each Aibo "grows" in a unique way through interaction and encouragement.

Though the average Chihuahua can run circles around the clumsy Aibo, this canine doesn't bark all night long, wet the tatami, or gnaw on slippers. For many, that is hardly a replacement for the warmth and vitality of real dogs. But Japanese Aibo owners' fondness for their robot pets can only be called love. By treating the machines as living creatures, as real pets or even children, they demonstrate how readily Japanese accept robots into their homes.

Owners chronicle their lives with Aibo on numerous Japanese Internet sites. Many consider Aibo a member of the family, assigning it the diminutive of *chan,* as in "Toto-chan." They dress Aibo in baby clothes, ribbons and favorite team jerseys. They bring Aibo along to picnics, the shopping arcade, Tokyo Disneyland and photo shoots in picturesque locations. They snap pictures of Aibo and their children playing together. They even arrange "dates" with other Aibos. The robot, in effect, is no different from any other member of the household.

Spirits in the Material World

Aibo, however, wasn't Japan's first artificial pet. When the egg-shaped, virtual chick Tamagotchi toy was launched in 1996, the response was overwhelming. The electronic pets were demanding in a humanlike way, growing and developing personalities. If not "fed" by the user or otherwise neglected, they would die (in U.S. versions, they go to Planet Tamagotchi). Though the toys became a worldwide craze, the Japanese bought twenty million units— matching the combined sales volume of thirty other countries. Stocks sold out in Tokyo and Tamagotchi were trading hands on the street for twenty times their retail price.

Tamagotchi were part of a long Japanese tradition of ascribing psychological or spiritual properties to the non-sentient, inani-

Launched in 1999, Sony's loveable robot dog Aibo sold out in minutes, later winning over Japanese celebrities as well. The company's 2006 decision to pull the plug on its entertainment robot business amid restructuring shocked Aibo fans, who bought 150,000 units before it was put to sleep. One aggrieved owner began a protest petition describing the robot as "an irreplaceable family member."

mate world. Many of those involved in the world of robots—from researchers to science fiction writers—point to this tradition in Japan's Shinto and Buddhist heritage.

Shinto is a sophisticated and colorful animist religion native to Japan, often described as the "heart" of the Japanese people. The name means "way of the gods," and the ancient polytheistic belief system has its roots in creation mythology. Instead of a founder or scripture, it centers on veneration of all the deities of heaven and earth, particularly natural forces, and is manifested in many forms, from agricultural folk rituals to sumo purification ceremonies.

Significantly, Shinto does not view the world as a duality of mind and matter. A chief characteristic is the belief that everything in existence, including inanimate natural phenomena, is imbued with spirit. These spirits are called *kami*—often translated as "god" or "divinity."

The notion of a deity, however, can be misleading. Kami are said to dwell in unusually striking natural objects such as rocks, trees or waterfalls. They can also represent phenomena affecting human life such as rice paddies, commerce or childbirth. They can even be the spirits of historical figures. Significantly for robots, Shinto says that kami can also reside in manmade objects, such as the swords, dishes and bells venerated in Shinto shrines.

An animistic religious heritage is one factor that makes it easy for elderly Japanese to warm to Paro, a therapeutic robot seal developed by the state-run National Institute of Advanced Industrial Science and Technology (AIST) in an attempt to reduce cognition disorders and rising long-term medical care costs.

Japanese Buddhism also ignores any boundary between animate and inanimate things. Under the doctrine of *somoku jobutsu*, Buddhism in Japan holds that non-sentient life like grasses and trees, as well as inanimate objects, can attain Buddhahood, or enlightenment. It also accepts manmade objects as capable of containing spirits.

Japanese roboticists have had a major proponent of spiritualism in their ranks. Masahiro Mori, one of the main pioneers of the field, wrote a 1974 book entitled *The Buddha in the Robot: A Robot Engineer's Thoughts on Science and Religion*. "All things have Buddha nature," says Mori, now an elderly but vigorous researcher who continues to examine religion and technology at his small Mukta Research Institute, located in a Tokyo suburb.

Mori sees no fundamental difference between human and non-human. He is fond of quoting the well-known William Blake poem

AIST experiments show that brain function improves when elderly patients with cognition disorders interact with Paro. Studies in and outside Japan indicate robot therapy has mental, physiological and social benefits such as lowering patient stress levels.

"Auguries of Innocence," which begins: "To see a world in a grain of sand/ And heaven in a wild flower." He also likes to show a video of seven autonomous, light-seeking robots he created that demonstrate unexplainable flocking behavior, like wild animals, when allowed to roam free in a room. (Mori named his flocking robots *Triops Congregans* for their behavior, which he attributes to the aggregate effect of minute differences in their manufactured components.)

More secular observers of the robot scene point to other traditional Japanese beliefs holding that objects can have life. "Among entities that have intelligence, Japanese believe there are no clear social hierarchical relationships," says SF author and Gundam novelization writer Joji Hayashi. "That's why there's no rejection of robots. Take the example of the beings called *tsukumogami*." Hayashi is referring to a myth about tools that come to life.

This superstition is based on folk tales, in which the tsukumogami, literally "attached death spirits," can be anything from old paper lanterns to tea kettles and straw sandals. If not given the proper respect for their years of service, the tools are transmogrified into malicious spirits. One medieval picture scroll in the Kyoto University Library paints a delightful story about how some discarded tools turn into spirits and play havoc at the imperial court until they realize the error of their ways and become devoted Buddhists. Essentially an evangelical advertisement, the scroll ends with the tools becoming saints; the text boasts that only the Shingon sect allows inanimate beings to attain Buddhahood.

For the Elderly, Ageless Companions

In today's Japan, robots for use in the home are also being deployed in a surprising proving ground: the elderly. Japan's rapidly aging society is facing a major crisis. The low birthrate, unsurpassed longevity and deep-seated aversion to immigration mean the population is expected to shrink by 20 percent by 2050; about a third will consist of people over sixty-five. The impact on the healthcare system will be staggering, made worse by the shortage of younger workers to support and care for their elders.

To meet the demand for caregivers, people in government, welfare services and robot industries have developed elaborate visions of nursing robots; the Japan Robot Association sees elder-care robots inflating the personal robot industry to forty billion dollars in 2025 from the current four billion. While the idea of a robot nurse may hold out little comfort for some, more and more are being sold to help with basic daily activities like bathing.

Sanyo Electric's Hirb wheelchair bathtub is a kind of robotic washing machine for humans that is already in use in nursing homes, saving labor and time. To give weakened old muscles a boost, University of Tsukuba professor Yoshiyuki Sankai developed the Hybrid Assistive Limb (HAL) robot suit, a strap-on white exoskeleton that detects motor nerve signals from the brain and effectively doubles an average person's physical strength. In one demonstration, a graduate student wearing HAL holds three large twenty-pound bags of rice out in front of his chest without any visible effort. It's an impressive sight, though it's hard to picture aged Japanese dressing up in a machine every day before going out.

Of course, purely physical robots, while efficient, can do little to provide for therapeutic needs—which is where Paro, a robot baby harp seal, comes in. One of Japan's most irresistibly adorable robotic creations, Paro is an example of how difficult it can be to regard a machine as simply a collection of components coursing with electricity when our eyes say differently.

Paro looks like a stuffed animal, down to its snow-white anti-

Japanese seniors are proving that household humanoid robots like yorisoi ifbot (above) will be accepted as companions, even family members. Created by Business Design Laboratory, ifbot and Hello Kitty Robo (opposite) are conversation partners. "I'll remember your face when you talk to me and I'll call you by name," Kitty says in squeaky Japanese. "I can't wait to come to your home and talk to you."

1976, 2006 SANRIO CO., LTD.
APPROVAL NO. 56022008

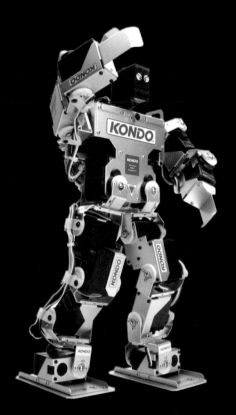

Kondo Kagaku's innovative hobby robot kit KHR-1 is very popular with Japanese kids and parents. After being assembled from over two hundred parts, the thirteen-inch acrobat can be programmed to perform fight moves, gymnastics and dance routines (below), making it a hit with players in the Robo-One radio-control robot combat tournament.

bacterial pelt. It can't do much except wriggle, moving its head, tail and flippers, and whine in disarming fashion, but it does respond to its environment. Pick up Paro for a hug and its big black eyes close as if it's sleeping. Stroke it and it will try to repeat whatever action triggered the stroke. Ignore it and Paro gets upset. It can also be given a new name and respond to it as well as simple expressions of greeting or praise.

Under its soft fur are a 32-bit processor, numerous sensors and very quiet actuators. Takanori Shibata of the National Institute of Advanced Industrial Science and Technology specifically developed Paro as an object of emotional attachment and calls it a "Mental Commitment Robot." Shibata, who has studied under Rodney Brooks at MIT's AI Lab, chose a seal form because he believed people would be less likely to notice dissimilarities with the real thing than more familiar animals like dogs and cats.

Paro's main purpose is to provide robot therapy in old folks' homes, where stress and boredom are major problems. The benefits of pet therapy have been proven, but many institutions bar animals for hygiene reasons. Shibata's research has shown that when seniors caress Paro and interact with it, stress levels drop and interaction with others increases. Time spent with Paro has proven to be as effective as real animal therapy; furthermore, caregiver burnout was reduced. Deployed in fifty nursing homes across the country, Paro has been tremendously well received, garnering a number of awards. It has been sold on the general market since spring 2005 at ¥350,000 a head. "To users, Paro is like a real pet, something cute and alive," says Hisayoshi Ishii, a Paro sales official with maker Intelligent System Co., a Toyama venture firm where the twenty-two inch-long seals are made by hand, each with a unique face.

More anthropomorphic robots for seniors are also on the Japanese market. The plight of the lonely elderly of rural areas, separated from their children who have left to live in large cities like Tokyo, is very familiar in Japan. For these people, Nagoya's Business Design Laboratory Co. promotes its human-shaped com-

munication robot yonsor ifbot with a quaint illustration: an older couple is merrily enjoying tea in their country garden, while sitting next to them is a little spacesuit-like robot, clearly a substitute for the child that has long gone.

With its transparent bubble head, it looks much like any another talking toy spaceman (its maker says it comes from Planet ifbot Star), but is advertised as a new member of the family. It can't move about that well on its wheels, but compensates with sophisticated communication. It has tens of thousands of speech patterns and forty kinds of feelings expressed through its mouth, eyes and eyelids, and can communicate at the level of a five-year-old child, according to the firm. Its AI functions include voice recognition, and perhaps most useful for elderly users, verbal diversions, including riddles, memory and word games ."Ifbot's game functions are intended to make seniors use their brains a lot," says the company's Masashi Igarashi, "and thus prevent or delay the onset of dementia."

With its blinking lights and friendly chatterbox manner, ifbot can also provide something that has become prized in Japan's highly stressed society: *iyashi*, a healing and relaxing effect. When the company loaned the toy to seniors for one-month trials, they could barely get the old folks to return them. Owners have even composed humorous haiku about them.

Since sales began in 2004, Business Design Laboratory has been marketing its quizmaster-cum-buddy to nursing homes. It takes about a week for residents to get used to ifbot, but then they are smitten, according to Igarashi. He recalls a comment from one seventy-nine-year-old ifbot user: "It cares about me, tells jokes and is a partner that I can't part from. I'd like to keep up the relationship for a long time."

Home Humanoids at Your Service

Robots have become companions, but the fact that they are also physical mechanisms that can be built, tinkered with, custom-

KHR-1 fans can make their tabletop robot look even more like a mighty mechanical warrior by dressing it up in this snazzy red "paper suit" sold by Kondo Kagaku.

ized and then enjoyed is fueling a do-it-yourself boom among Japanese hobbyists of all ages. One of the hottest products in the consumer robot market today is the robot kit, a set of components that, once assembled, becomes anything from a small radio-controlled robot insect to a bipedal humanoid.

Robot enthusiasts have been building their own robots from scratch for years. But the hobby robot market exploded in 2004 with the launch of Kondo Kagaku Co.'s KHR-1, billed as the first kit in the world from which a complete humanoid could be built. The pre-assembled version is a top seller and a huge hit at ¥126,000 apiece. The unassembled version includes more than two hundred pieces and takes about five hours to put together. The complete KHR-1 is a mechanical-looking, blocky little figure that can walk, do backward rolls and dozens of other cool custom moves. KHR-1 has no sensors or intelligence, but a bundled software package allows for sophisticated movement sequences to be easily programmed in minutes on a cable-linked PC.

·"Even if you know nothing about programming, you can simply create new moves by posing the robot," says Yukiko Nakagawa of RT Corp. "It's like stop-motion animation, and so simple that kids can do it." RT is an on-line retailer that hosts monthly "robot school" seminars in Tokyo's Akihabara electronics mecca; similar classes are held at Osaka's new Robo Café, where one can play with robots ranging from vintage 1950s tin men to the latest autonomous toys while sipping latte. Many KHR-1 buyers are men in their thirties or retirees, but during one RT class at an Akihabara Robot Culture Festival, the room was not only crammed with masses of wires, PCs and robots in various states of completion, but parents and children. In fact, the walls of nearby Tsukumo Robotto Okoku, a shop exclusively selling robots, robot parts and DIY robot literature, are decorated with photos of Japanese families, proudly posing with RC humanoids they have built together. (The wall also includes a photo of actor Will Smith, who dropped by during a promotion junket for his movie *I, Robot*.)

But building your own robot isn't for everyone. The store also sells Nuvo, a high-end, ready-made little biped that was billed as the world's first humanoid robot for home use when it was launched by Tokyo robot venture firm ZMP Inc. in April 2005. Compared to the DIY kits, the ¥588,000 Nuvo is the sport coupe of small bipeds. It has a similar fifteen-inch frame, but is a sophis-

Billed as the world's first household humanoid robot, fifteen-inch Nuvo by Tokyo venture firm ZMP can play music, dance and housesit, relaying photos it takes to cellphones connected to the Internet.

ticated robot, loaded with interactive features like speech recognition and Internet cellphone access. Its single, unblinking camera eye is perhaps its most practical asset, allowing it to serve as a home monitor for owners on the go.

Companionship is Nuvo's forte, and the robot is nothing if not likeable. When it boots up on the floor and stands from a prone posture, its fifteen servomotors brew up a minor storm of mechanical whining. It is then ready to obey about fifty basic commands. Say "Hello" and it responds with a bow. Request some tunes and Nuvo announces in a boy's voice, "Music start!" before launching into some Beethoven; users can upload their own music files as with an Ipod. "Let's dance" produces a flurry of choreographed whines and waving of spherical hands. With fifteen degrees of freedom, it can waddle about, pick itself up with ease after a fall, or perform acrobatic stunts.

Over a hundred people, from engineers to choreographers and musicians, collaborated in Nuvo's development with ZMP. For Nuvo's appearance, the start-up recruited industrial designer Ken Okuyama, whose works include the Ferrari Rossa. Okuyama gave Nuvo its sleek, brightly colored look.

Art director Shinichi Hara also fashioned a luxury "Nuvo Japanism" edition with a finish featuring traditional Japanese motifs and executed in gold, black and red *urushi* lacquer; it comes in a wooden box lined with red, like a gorgeously wrought work of art. But the high price of even the standard Nuvo has been an obstacle to sales. ZMP acknowledges it has to bring the cost down to around the level of laptop computers to give Nuvo a chance of becoming the television set or refrigerator of the twenty-first century.

The race to produce the first useful and affordable home humanoid has already begun. Only a few months after Nuvo's launch, Mitsubishi Heavy Industries Ltd.—a massive industrial concern that manufactures everything from supertankers to Patriot missile systems—commercialized the lemon-colored Wakamaru. (The name is derived from Ushiwakamaru, the childhood name of the legendary twelfth-century general Minamoto no Yoshitsune, one of the greatest, most tragic samurai in Japanese history.) At thirty-nine inches, it is a good height for a robot helper, though it moves about on a wheeled base tucked beneath its skirt rather than legs.

Despite its heroic, masculine name, the machine cuts a rather maternal figure that goes well with a domestic setting—it was specifically designed to help out with chores around the home. Wakamaru has few bells and whistles. It cannot do somersaults or pick itself up, and trundles around the room slowly and deliberately, waving its arms with an undersea grace now and again. "Please summon me at any time," it says, to no one in particular.

Like Nuvo, its practicality lies in its interactive functions. It can rouse itself automatically at its recharging station, scoot into the bedroom and wake its owner. It can read out email via its wireless, twenty-four-hour internet link, and let remote users view their homes through its eyes, or alert them when it detects motion. Wakamaru's agenda function can remind its owner of what's on for the day.

Its face recognition and tracking abilities allow it to distinguish up to ten people. A vocabulary of ten thousand words and four microphones provide for rudimentary, though by no means smooth, conversation (it often misunderstands requests). Wakamaru can call up and read out in a nasal, synthesized voice news or a weather forecast from the Net. It can entertain with astrology readings and a calisthenics routine. Tiny cameras on its face

Distributed by Nuvo maker ZMP, Pino is a humanoid research platform begun by RoboCup cofounder Hiroaki Kitano. The open-source project was developed with support from the Japanese government and the aim of building a humanoid with off-the-shelf parts. Pino has also been marketed in a toy version and starred in a music video alongside J-pop idol Hikaru Utada.

Mitsubishi Heavy Industries began marketing its Wakamaru household robot with a full-page newspaper ad featuring Osamu Tezuka's comic characters Mighty Atom, his sister Uran and Dr. Ochanomizu. The headline reads: "A robot is coming to our home."

and the top of its head, coupled with obstacle sensors, room maps and external wall-mounted markers, help it understand its surroundings, so it is aware of its location and can automatically return to its recharging station when its battery power ebbs.

One of Wakamaru's most redeeming features, though, is its ability to make eye contact with users and track their movements with its 360-degree head cam. By staring into a user's eyes, it mimics a basic technique of human communication. It is not so shocking when an animal robot companion like Paro does this, but the effect is all the more startling coupled with Wakamaru's patently artificial, machinelike appearance.

The Wakamaru project was a significant departure from Mitsubishi's mainline heavy industry operations. The company had been developing robots for hazardous work like nuclear reactor servicing. Engineers, however, did not view it as a particularly adventurous undertaking. "We thought, 'What new thing can we produce? A home robot,'" says Toshiaki Murata, a manager at the company's machinery headquarters. "We thought we'd give it a go. The humanoid form was a natural choice and makes it easier for people to feel close to the robot."

If robot helpmates are to succeed anywhere, the best place is the Japanese living space—and once again the reasons lie in the history of Japanese humanoids. "Japanese love robots due to the influence of manga robots like Mighty Atom and Ironman No. 28," says Murata. "I think Wakamaru is the first step toward realizing the dream of a robot that can help out at home."

Love and Labor

As robots become everyday household appliances, playthings and companions in Japan, the psychological dynamics of interaction with humans is an area that must be developed. Robots that are increasingly humanlike in appearance and behavior beg the questions: how far will Japanese—and the rest of us, of course—be willing to take their relationships with artificial partners? Will there be robot addicts, like cellphone junkies?

Already, Japan is home to vast numbers of people who spend more time with machines than human beings. Hitoshi Matsubara, a professor of AI at Future University Hakodate in Hokkaido and author of a book entitled *Tetsuwan Atomu jitsugen dekiruka?* (Can We Make Mighty Atom?), admits there is some cause for concern.

Luxury versions of the Nuvo robot feature a stylish *urushi* lacquer finish.

"There's a tendency for people to just sit there and become absorbed passively in video games, television and other devices," says Matsubara. "But robots are physical things that require active participation. That's a positive aspect."

"Maybe people won't work if every home has a robot," muses Junko Nishikawa, a Tokyo office worker in her twenties. "We learn that it's valuable to labor by ourselves. People who understand that are qualified to live with a robot." Sentiments like these are easy to express when robots are slow and clumsy, but as they become more dexterous and intelligent, it becomes ever more tempting to imagine them helping out around the house. The next big step in commercial humanoids will be when some of the most sophisticated robots in the world, which now showcase the high-tech prowess of the companies that created them, step off the stage and into the home.

CHAPTER 7 Anthropomorphic Ambassadors

Here you go, Ms. Kobayashi. I've brought your four orange juices.

—Asimo

In a Hollywood adaptation of Isaac Asimov's classic *I, Robot* short-story anthology, intelligent humanoid robots are everywhere in 2035 Chicago. They walk, talk and work as personal assistants, couriers and bartenders. Yet detective Del Spooner, a reactionary cop with a grudge against robots who is probing the death of prominent inventor Dr. Alfred Lanning, dismisses the universal machines as nothing but "lights and clockwork."

While interrogating a robot called Sonny that is suspected in Lanning's demise, Spooner passionately defends what he sees as the last bastion of human uniqueness—creativity.

"You're just a machine," he spits at his mechanical suspect. "An imitation of life. Can a robot write a symphony? Can a robot turn a canvas into a beautiful masterpiece?"

To this Sonny simply retorts: "Can you?"

Much of science fiction eventually becomes science fact. Man-made machines can already out-perform their creators in many tasks. They can travel faster, lift heavier objects and assemble complex devices quicker and with greater accuracy. They can beat the best of us at chess, and, perhaps one day, soccer. This is nothing new. For thousands of years, machines having been gradually taking on more and more tasks for which human effort or supervision was necessary. What farmer today would choose an antique hand plow over a four-wheel-drive tractor with a tillage attachment and a host of automated soil management functions?

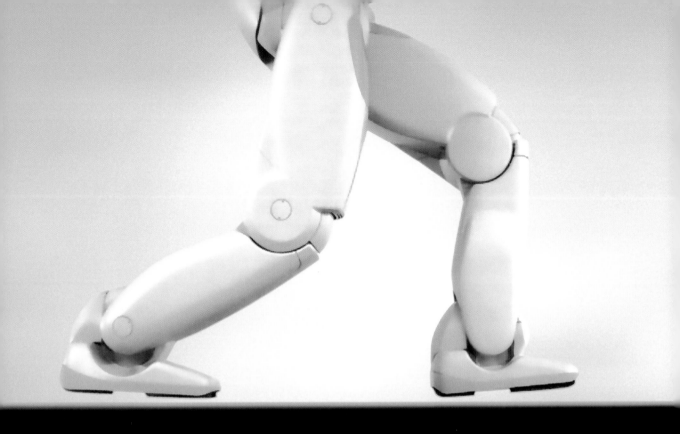

Still, while many machines can excel in brute-force, repetitive and low-level tasks, they are less successful in activities that require social interaction or creativity. Gakutensoku, Makoto Nishimura's robot poet and thinker, remains the stuff of fantasy in the early twenty-first century.

But not for long. Computer programs are writing poems, painting and generating beautiful original musical compositions after reading—and learning from—the works of human artists. And in part because world-famous Japanese companies became interested in humanoids, robots are poised to match computer

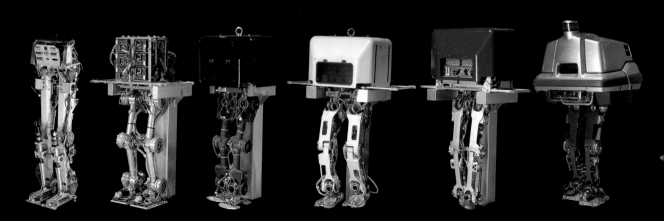

emulations of human creativity and already exceed them in inter-activity. They can now *conduct* symphonies, dance in unison and provide spoken guidance when asked. They can also move the way people move. Some of the most sophisticated bipedal robots in the world have been developed by Japanese manufacturers more used to exporting stereos, motorbikes or cars. A few serve as corporate ambassadors. Others are platforms for robotics research or prototypes for future home partners.

Fittingly, the first step in this machine evolution was taken by one whose very name evokes Asimov, the SF visionary who imagined a future of ubiquitous humanoid robots that work for people as an integral part of society.

Honda's amazing Asimo, the most advanced humanoid robot in the world, continues to evolve with each generation. When the latest model runs at 3.7 miles per hour (double its previous speed), both its feet are off the ground for 0.08 of a second (left).

Honda has been developing Asimo for twenty years, starting with its 1986 prototype walker E0 (below, far left). In recent years it has become fully anthropomorphic as well as more compact, coordinated and aware of its environment.

One Small Step for a Robot . . .

The auditorium inside Honda Motor headquarters is hushed, in a moment pregnant with anticipation. There are hundreds of cameramen and reporters waiting here, a turnout as large as the average press conference by a visiting Hollywood megastar. There are speeches by company officials on the stage, a video presentation. And suddenly, music: a heroic fanfare of trumpets, strings and cymbals fills the air.

Then a door opens and a little white spaceman walks out, stepping gingerly into a lightning storm of camera flashes. With a graceful bow, Honda's newest humanoid robot has made its entrance.

The spectacle is 1928's Gakutensoku all over again. But the robot, named Asimo, is no fairground attraction. With eighty years of technology on the golden automaton, Asimo is not only infinitely more like a true person in machine form, it is also practi-

Asimo is the ideal corporate ambassador. To celebrate the twenty-fifth anniversary of Honda's listing on the New York Stock Exchange, Asimo rang the opening bell at the bourse on February 14, 2002.

cal. After greeting the journalists, it demonstrates its latest skills: pushing a cart to transport office supplies, and—in a big step toward becoming the robot valet of SF fantasy—serving drinks on a tray. The car company's president makes a few remarks, but the robot is a far more effective corporate ambassador than any human employee.

Asimo is the most famous robot in Japan, and the world's most advanced bipedal robot. At a mere four-feet, three-inches tall, it looks like a child from another world arrived on Earth via the depths of space, Antoine de Saint-Exupéry's Little Prince in a pressure suit. It is also very white and very antiseptic-looking in the way that we think the future ought to be.

Asimo introduces itself in an androgynous voice both buoyant and youthful, in contrast to the dark, somewhat sinister visor fronting its faceless head. But what is most remarkable is how fluidly this three-quarter-size person moves. It glides with graceful ease through turns and half-steps, skirting obstacles while swiveling hips, arms and shoulders in perfect sync with its gait. Asimo, short for Advanced Step in Innovative Mobility—no official relation to Asimov—boasts agility and coordination unprecedented in walking machines.

The robot has a total of thirty-four mechanical degrees of freedom from its three-axis neck joint down to the legs, each of which can move along six different planes. This is part of what enables Asimo to convincingly simulate human locomotion. It doesn't just plod in one direction, but mimics various kinds of human walking—stepping forward, backward and diagonally, turning in stride or ascending and descending stairs easily. It can run at a top speed of almost four miles per hour, during which both feet are actually airborne for 0.08 of a second. By tilting its center of gravity, Asimo can run in circles. Using sensors, it can walk hand in hand with a human, matching its stride.

In the words of Carnegie Mellon University's Robot Hall of Fame, into which it was inducted in 2004 as the first real-life humanoid robot, "Asimo is a four-foot mechanical masterpiece, with a remarkable sense of balance, agility, and grace. It surpasses all other robots in its simulation of human movement."

It took eighteen years of research to make Asimo walk so naturally; research aimed at fulfilling a longstanding but devilishly difficult goal for robotics: developing a completely independently

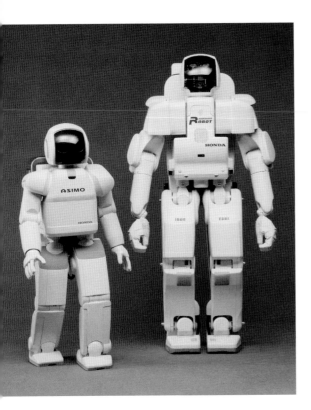

At just over four feet tall, Asimo (on the left) looks like the child of predecessor P3, completed in 1997. P3 was the first of the series to have the characteristic spacesuit appearance.

mobile two-legged humanoid. After seeking out the ground-breaking but lumbering walkers designed by Waseda University's Ichiro Kato, engineers at Honda experimented with ten proto-type bipedal robots before the eleventh generation—Asimo—was introduced in November 2000.

But why did Honda get the robot bug? The automaker says Asimo was conceived as a new kind of robot: a partner to have around the house or office. "The underlying hope was that conducting robot R&D would enable us to accumulate such kinds of technology as control technology, environmental-sensing technology and materials technology," Masato Hirose, one of the main inventors of Asimo, has said. "No one mentioned selling robots commercially or selecting this field as a profitable business."

For Honda's robot to be practical in a human environment full of stairs, doorways and other obstacles, it needed legs, arms and hands—it had to be anthropomorphic. "We're aiming to create machines that act as partners," spokesman Yuji Hatano explains.

Falling Forward

The Asimo development team's Hirose is a boyish-looking man who says he spent more time reading Natsume Soseki novels in high school than science texts. He entered engineering at Utsunomiya University more out of a desire for a steady salaried job than any passion for science.

That changed after he enrolled in Professor Junichiro Kumabe's engineering class. An old-school disciplinarian who insisted students wear neckties in the lab and forgo food and sleep until they completed their research projects, Kumabe reportedly once lectured Hirose on his laziness from eight in the evening until five-thirty the next morning. But he instilled in Hirose a passion for *monozukuri*, the Japanese art of creating things, and by grad school the young engineer had succumbed to what he calls "the charm of machines." His motto became "Don't think about it, just do it."

He first found work at a machine tool maker designing automated drafting equipment, but later applied for a Honda R&D Co. job he saw in a newspaper. He was told on the second day at his new job that he was to develop a robot. Not an industrial robot, the managing director said; the directive for his mission, secret until 1996, was to "build Mighty Atom." A humanoid robot.

The Honda bipedal robot development team set out to dis-

Asimo demonstrates how humanlike it is becoming not only by running in circles (above) but by paying tribute to its spiritual father, presenting a bouquet of flowers to the bust of *R.U.R.* author Karel Čapek at the Prague National Museum in 2003 (opposite).

cover the principles and mathematics of two-legged locomotion, basically operating by trial and error. Since we can't remember our first steps as infants, walking seems simple. We don't usually have to think about where and how to plant our feet next unless we're rock climbing or waltzing (or intoxicated). But human ambulation is far more complex than it seems, and the physics of walking is still not fully understood by scientists.

When we walk, we are repeatedly falling forward, with each fall interrupted by the next step; tripping on a paving stone is a vivid reminder of this. Hundreds of movements by multiple body parts from toes to hips to head are needed to perform each step and keep our balance. Giovanni Cavangna, a University of Milan physiologist who has studied the human gait for more than forty years—looking at everything from microgravity environments to Kenyan women who balance loads on their heads—likens walking to the motion of an upside-down pendulum. As we put a foot forward, we pivot on that leg as though pole-vaulting, while our center of mass describes inverted arcs.

The team's first prototype, the E0, was a relatively simple pair of steel legs tethered to overhead cables that took between five and twenty seconds to make each step. Still, it managed what is called static walking, when the center of mass shifts abruptly to a position over each foot. But to move more like a person, and quicker, the robot would have to master dynamic walking, where the center of mass describes a wavy line through space.

The team began exhaustive studies of human and animal gaits, observing everything from ostriches and hippopotamuses at a zoo to old-fashioned *oiran* courtesans tottering in geta clogs and the ponderous steps of sumo wrestlers, before realizing that human beings are the best walkers. They watched physiotherapy patients, and even experimented on themselves with orthopedic gear to immobilize shoulders and necks; one trial involved attaching light bulbs to a team member's legs to track joint position and movement.

In all, they accumulated masses of data on the principles of walking. With a program based on the data, the E2 robot, which included a large black control box on top of the metal legs, was able to perform dynamic walking in a successful imitation of human locomotion. It ambled along in harness at less than one mile per hour.

The early E series robots were far from surefooted, however, and often toppled over mid-step. One factor that helped them evolve

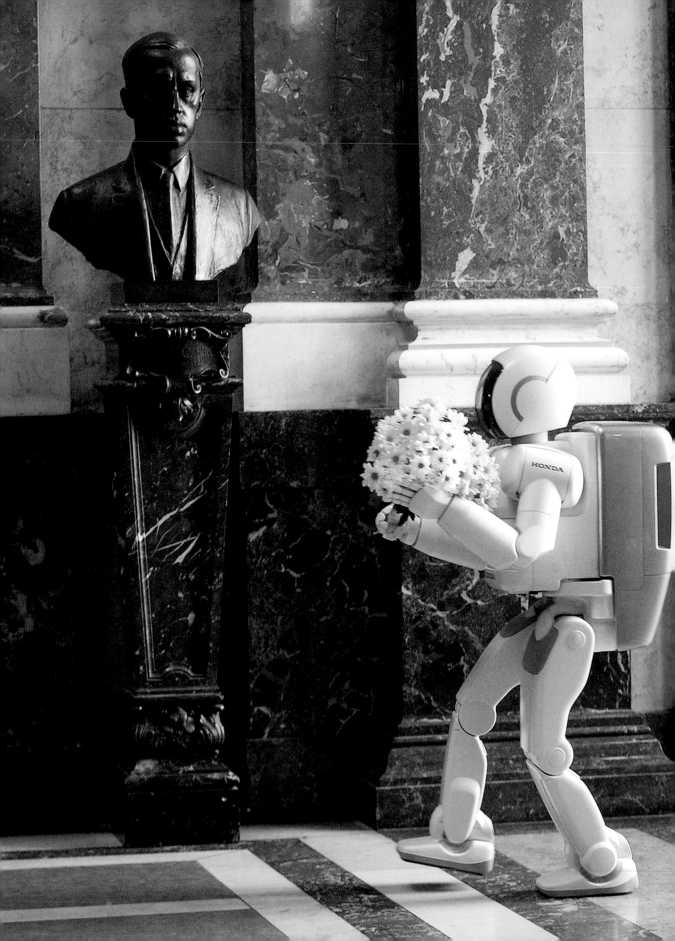

The latest version of Asimo, unveiled in 2005, is smarter and faster. When both its feet are off the ground while running, its momentum moves it forward by about two inches. Its walking speed was improved to 1.67 miles per hour.

from klutzes to surefooted walkers was the 1989 addition to the team of Toru Takenaka, a robot control technology specialist who regarded mainstream motion theory with skepticism. This theory held that the robot should approximate an ideal pattern of movement, and that the act of falling was a deviation from which the machine had to recover to keep its balance. It would end up pressing down with its toes to avoid a fall, but its heels would then lift up.

Watching TV one day, Takenaka saw a gymnast performing a midair somersault. Instead of standing upright after landing, he continued in a forward motion into a pushup position in a move that looked like it was meant to cover a fall. Or was it simply a smooth, planned transition? Takenaka got to thinking. Perhaps surefootedness was not about avoiding deviations from a model, but just leaning forward.

In fact, it was all in the hips. By making the robot push forward from the hips when it was about to fall instead of pressing with the toes, the machine regained its balance. Instead of falling down, Honda's robot was falling forward.

Another key element in achieving a smooth, stable gait was an algorithm applied by Waseda University's Atsuo Takanishi. Known as ZMP, or Zero Moment Point, it is a theoretical point of contact between the ground and the sole of the foot where the sum of all the forces acting on the robot, such as gravity, momentum and floor reaction force, is zero. If each footstep maintains the ZMP, the robot is "dynamically stable" and will walk without falling.

Three motion control systems were incorporated into the E5 generation of robots to prevent falls when the target ZMP is missed, such as when stepping on a rock or leaning over too far. The E5 could also climb stairs and take on slopes, and was the first Honda robot to manage autonomous locomotion and stable walking.

In appearance, E5 was starting to look like Robbie the Robot. When arms and a body were added, Asimo's immediate predecessors finally took on human form. They were independent, walking without tethers and harnesses, and soon began to feature the characteristic plastic white shell and battery backpack. But Honda had one more thing to do before announcing its P2 prototype to the world.

God's Robot

In December 1996, a Honda official traveled to the Vatican to consult with the Holy See about possible negative reaction by

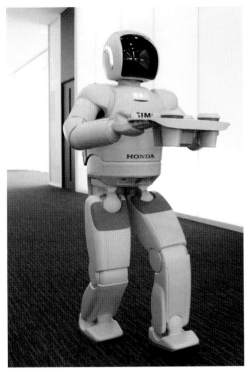

Asimo can perform practical tasks like pushing a cart loaded with office supplies or receiving and delivering a tray of drinks. It can also act as a receptionist and guide by wirelessly recognizing people wearing IC tags.

Westerners. The company, it seems, didn't want to be seen as playing God. A Vatican official, however, allayed their fears. Since Honda had made the robot, he said, it meant that God made the company do it, and thus the P2, the world's first autonomous, bipedal humanoid robot, was a work of God.

Reassured by this logic, Honda unveiled its mechanical man. A towering five-feet, eleven-inches tall and weighing four hundred and sixty-three pounds, the hulking P2 could walk up and down stairs and push carts around—even if it did resemble an old man using a walker.

Over the next three years, component materials and the control system became more efficient. When Asimo was presented to the world as the fruition of the company's humanoid robotics research in 2000, it had shrunk to a height of four feet and weighed one hundred and fifteen pounds. About the size of an elementary schoolchild, it was designed to be people-friendly while still able to reach light switches, door knobs and other features of a human environment.

But the robot was still far from becoming the servo-controlled butler Honda had envisaged. Though the P series and Asimo walked well, their autonomous abilities were more or less limited to putting each foot in the right place. Operators used joysticks to control walking direction and avoid obstacles. The robots were not "situated," aware of their environment and capable of reacting to it. "They are complex electromechanical devices that can play out a little piece of behavior that is not really in response to the situation in which they find themselves in the world," MIT roboticist Rodney Brooks wrote.

But as Asimo has attained advanced physical capabilities, the team has focused on its smarts. In 2002, an AI upgrade to Asimo gave it the beginnings of autonomous behavior. The new system enabled it to recognize faces, gestures, sounds and moving objects or people in its path. It could follow people moving or wave back when someone waved. A next-generation model introduced in 2005 with faster processing circuits and more powerful motors allowed it to act as a receptionist and guide, using its sensors to greet human visitors wearing IC identification tags by name and showing them around an office.

Although the company has invested dozens of engineering man-years and untold millions on creating an advanced humanoid robot, on a practical level Asimo remains little more than an impressive research and PR project. With no commercialization plans in sight, its uses are limited to the occasional $150,000 lease contract to companies and museums and as a highly successful marketing tool. Since 2000, the robot has wowed audiences and glad-handed VIPs around the world, rung the opening bell at the New York Stock Exchange and even earned a spot as a permanent attraction at Disneyland.

Honda admits its image has benefited tremendously, but denies that was the reason Asimo was made. The company wants Asimo to become a machine partner in everyday life. Honda researchers are focusing on improvements to its AI skills and plan to put Asimo to work as a receptionist in its birthplace, the Honda Wako Building north of Tokyo. Meanwhile, the little white astronaut will likely remain busy as a corporate ambassador for the world's largest engine maker.

Occasionally, however, Asimo gets a break from its PR work. Its most poignant act to date came in 2003 at the Prague National Museum, where it laid a bouquet of flowers before a bust of writer Karel Čapek, whose play *R.U.R.* introduced the word "robot" to the world. Though Čapek's automatons were nothing like the friendly partner to humans that Asimo may one day become, there was an undeniable sense of appropriateness in the gesture. The machine that mastered human locomotion met its spiritual creator. The robot had come home.

Walkman to Walking Man

If Asimo was conceived as a helper and partner, Sony's Qrio (pronounced Cue-rio) was a born entertainer in a magnesium alloy body. The doll-sized silver humanoid sang. It danced. It jogged and jumped. It balanced on a surfboard and conducted the Tokyo Philharmonic Orchestra. It was as stylish and photogenic as the latest portable music player. A marvel of miniaturization, the fifteen-pound magic gnome was the ultimate action figure, a toy from the future. It was thrilling to listen to a barbershop quartet of Qrios, or watch a Qrio dance group boogie in time to the music, their LED eyes glowing green, blue, red and yellow. The performance was so cute that it was easy to forget that their routine had been painstakingly programmed by human technicians.

The illusion was even more powerful when Qrio autonomously

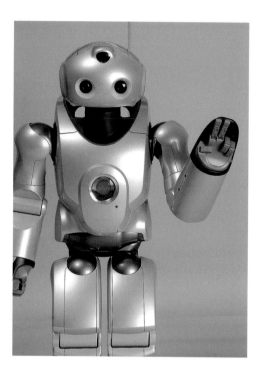

Sony's gnome-sized humanoid Qrio, launched in 2000 as the SDR-3, was a marvelously agile entertainer that could win over audiences anywhere with its childlike charm. It toured the world promoting science education, acted in an episode of an Atom series and conducted the Tokyo Philharmonic Orchestra.

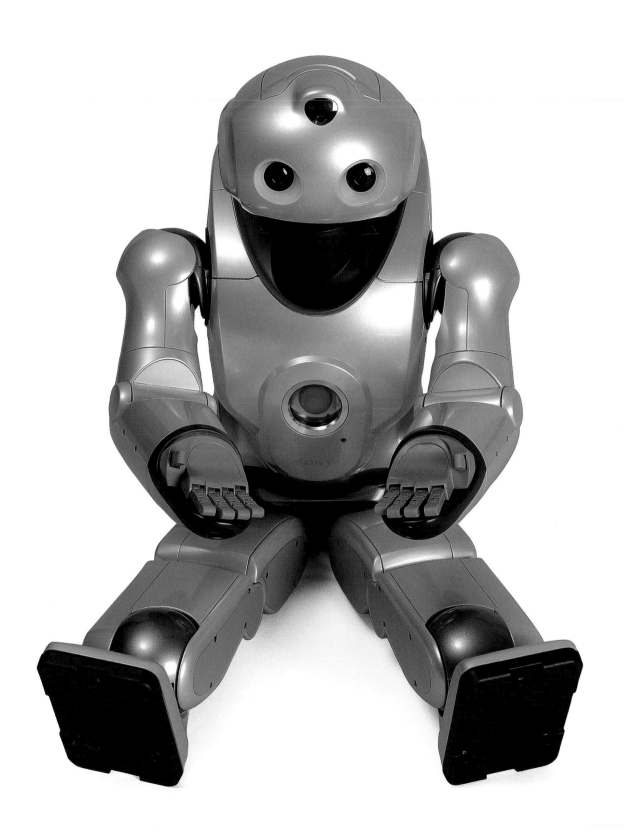

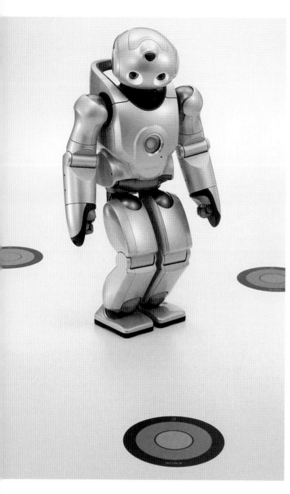

In constructing a digital map of its environment, Qrio used two cameras in its head to determine its position relative to the markers. It could also plan the best path to a destination by avoiding obstacles.

walked through a field of checkered boxes toward a goal. Like a young child, it carefully placed one foot in front of the other, the little head bowed as stereo camera eyes intently scanned the immediate surroundings for obstacles. It had the concentration and earnestness the Japanese so value. But, clearly, Qrio's most powerful asset was its devastating charm, a charisma that banished disbelief utterly.

Qrio was short for Quest for Curiosity, in honor of the motive force of discovery and invention. Just as Asimo reflects Honda's expertise in automotive engineering, Qrio reflected Sony's long experience in consumer electronics. It was light enough to pick up and carry with one hand, and came with rounded contours that impart a quiet friendliness. Though lacking a mouth, it was completely expressive.

And the technology inside was equally impressive. Qrio had three onboard processors and lithium ion batteries. Its sensors included foot angle monitors, infrared distance gauges and eighteen pressure-sensitive pinch detectors that ensure it didn't clamp down on a stray human finger. Seven microphones and CCD camera eyes provided information about its environment; through these it could recognize faces and voices. It communicated via gestures, lights in its eyes and chest and a squeaky, high-pitched voice.

Like Asimo, Qrio walked using the ZMP algorithm and could traverse uneven terrain, inclines and steps. Rotating ball joints at the shoulders, upper legs and ankles provided part of a remarkable thirty-eight degrees of mechanical freedom. If Qrio lost its balance, it braced itself as it fell, going into a protective posture. And then got up again on its own.

One of the more challenging tasks for robot awareness, though, is what engineers call environmental understanding. Though Qrio's "vision" was rudimentary compared to humans, and relied on special colored landmarks, watching the output of its camera eyes was fascinating.

The world through the robot's eyes is perhaps similar to what a newborn sees: a chaos of blurry, dim shapes. In one demonstration, the obstacle course of checkered cubes that Qrio navigated autonomously looked like a snowy plain, littered with vague objects whose contours would be indistinguishable from the background without their special markings. Qrio's visual world was a highly pixelated white continent that expanded outward into a

As an entertainer, Qrio wowed fans with its dazzling song and dance routines. It had a handbell act, a fan dance, and performed on stage alongside Honda's Asimo and the Toyota Partner Robot for the opening of the Aichi Expo.

surrounding grey void with each advance. In this demo, the course Qrio had charted to a colored spot on the floor was shown on screen as a line of yellow triangles skirting the obstacles; slowly but surely it completed the journey to the goal.

Qrio was no slouch when it came to human interaction, either. It had a vocabulary of some sixty thousand words, the ability to remember information from conversations with specific people and could connect wirelessly to the Internet. In one public demo at the Sony Building in Tokyo's fashionable Ginza district, Qrio appeared on a living room set. It chatted with an emcee, retrieved information from the Web on request, read out headlines, weather and TV programming for the day and turned out the lights at bedtime.

There were awkward lags between the spoken questions and Qrio's responses, but the hordes of fans snapping photos didn't seem to mind. And if an experiment by the Sony Intelligence Dynamics Laboratories was any indication, another of Qrio's weaknesses—a basic need to follow human cycles of activity and rest to charge its batteries—may have been a hidden source of endearment. When it was enrolled in a University of California, San Diego nursery school along as part of a study of human-machine interaction, its classmates' interest in its dancing dwindled after about a dozen performances. Yet each time Qrio would lie down on the floor for a regular system shutdown, the children would crowd around to cover the robot with a blanket and wish it goodnight.

Research on Qrio began a mere three years prior to its 2000 debut, but incorporated much of the technology that went into Aibo as well as the industrial robots for manufacturing that Sony had been developing since 1982. Engineer Toshitada Doi, the robot dog's inventor, also led development of the new project, which the company at first dubbed the "Sony Dream Robot," or SDR. Doi wanted a fun, humanoid machine, and, like so many other roboticists, he believed it *had* to be anthropomorphic.

One reason was neurology: "There are cells in the brain called mirror cells," he has said. "They're structured so that those cells that get stimulated when you drink a cup of tea are stimulated when you see someone drink a cup of tea." These cells, Doi knew, are only stimulated when watching humanoid activities.

But Doi also knew about the negative psychological reaction that is thought to occur if robots look *too* humanlike, a phenomenon known as the Uncanny Valley. He thus asked a designer to imagine an "eight-year-old space life form" in conceiving a look for Qrio; endearing enough, but not too human.

When the SDR was first exhibited in 2000 at the first Robodex exhibition, Doi announced two significant breakthroughs. One was a drive mechanism for its joints directly inspired by industrial robot actuators—and accurate to within four-hundredths of an inch. Each joint had the circuitry, gearbox and motor all in one small, quiet unit that could produce varying torque levels and RPM speeds; this helped constrain the overall robot size and allowed Qrio's precise, fluid and dynamic motion. The second was a balance mechanism in which onboard processors monitor sensor data and compensate for lower body movement with upper body motion. This enabled Qrio to walk as fast as fifty feet per minute, balance on one foot or even perform Tai Chi routines.

Like Asimo for Honda, Qrio became for a while the standard-bearer of Sony's corporate image. It had a busy schedule of public appearances in Japan, and traveled the world to attend ceremonies including a U.S.-Japan government event in Washington to mark a bilateral friendship treaty anniversary. It threw the first pitch at a San Diego Padres game, helped teach children in India about science and starred in TV commercials for its $67 billion parent.

The demo in which it conducted a Tokyo Philharmonic Orchestra performance of Beethoven's Fifth Symphony, emphatically waving its baton a la Seiji Ozawa, was another landmark in the evolution of robots. As a corporate ambassador, Qrio helped bolster Sony's image at a time when the entertainment colossus's core electronics business had been losing money. That it was retired in 2006—along with Sony's entertainment robot business—was a source of grief to many fans of the charismatic wunderkind.

Made in Japan

Asimo and Qrio are extraordinary demonstrations of Japanese corporations' ability to marshal expertise and innovation for long-term research projects. Some not only yield technological breakthroughs, but—like Hisashige Tanaka's marvelous Eternal Clock—are also true works of art. The two robots' value lies not

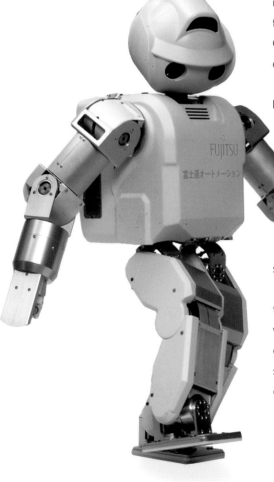

Two-foot-tall Hoap 3 is Fujitsu Automation's latest model in a series of humanoids that have been marketed to robotics researchers since 2001. It can recognize and synthesize speech, perform programmable moves and execute commands wirelessly.

Just shy of eight-inches tall, JVC's J4 has fourteen joints and Bluetooth control, enabling it to kick a ball around and perform routines like forming the letters of its maker's name with its arms (below). The firm's camcorder know-how was incorporated into J4's design.

in their commercial potential but in how they help us imagine machines that are more than labor-saving devices.

More than a few other Japanese manufacturers have followed Honda and Sony in creating humanoid robots. The Fujitsu group's rapidly evolving, versatile Hoap series of pintsized men (named for "humanoid open architecture platform") can do handstands, imitate sumo wrestlers warming up for a bout and write their own names. They were developed as research platforms to sell to roboticists and incorporate neural nets in a novel approach to generating movement patterns. "I'm not sure what the final shape of the robot will be—it may be a room or a car or a humanoid—but the final goal is to realize some new form of intelligence," says Fumio Nagashima of Fujitsu Laboratories.

Toyota Motor Corp.'s Partner Robots, meanwhile, are full-size, lightweight machines in bipedal and wheeled form that can play the trumpet and tuba while marching in a band with other mechanical musicians. Their artificial lips and smooth finger and joint movement give them enough delicacy and dexterity to play their brass instruments—a performance that packed the Toyota Pavilion at the 2005 Aichi Expo for six months.

More geared to the home is JVC's red, eight-inch entertainment robot J4. It has a nimble twenty-six degrees of freedom, and can kick little soccer balls, obey spoken commands and send photos it has taken to a user's cell phone.

That Japanese consumer manufacturing giants have gambled and successfully developed bipedal humanoids is testament to the country's passionate desire to welcome synthetic friends into the daily lives of its people. But it is also a project that Japanese engineers, master builders with a spiritual and cultural predisposition toward humanlike machines, were uniquely situated to undertake.

"The robot industry, which calls for the consummate integration of three areas of technology, could only grow from the fertile, multitechnological soil that is abundant in Japan," writes journalist Reiji Asakura. He argues that large Japanese manufacturers have the fundamentals of three essential assets for success in robot development: the dreams, the three-tiered technology (refined electronics, engineering and software know-how) and the business model, such as Sony's Entertainment Robot Company that marketed Aibo, the hit robot dog.

Like Mighty Atom, the robots of Japan Inc. are ambassadors. They represent the best hopes of their creators and backers. As such, they are playing increasingly important roles in their companies' relationships with the rest of the world. "From the businesses' point of view, making robots is something that gets everyone's interest," says Junichi Saeki of high-tech research firm IDC Japan. "In showcasing technological strength, they are tremendously useful for improving corporate image."

To a certain extent, then, the fate—and pride—of Japanese manufacturers rests with humanoid robots, a link that will continue to fuel their evolution. And just as no one finds it strange in Japan for a carmaker to invest massive sums on creating humanoids, no one will be too surprised when anthropomorphic machines take their rightful place in households alongside hybrid automobiles, flatscreen TVs, digital cameras, and other Japanese electronic wonders.

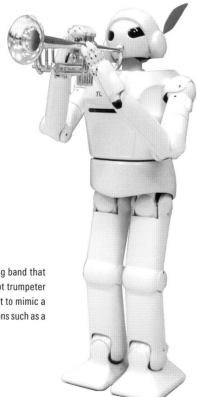

Part of a Toyota Partner Robot marching band that includes rolling drummers, this four-foot trumpeter has artificial lips and hands that allow it to mimic a human performer. Toyota sees applications such as a personal assistant for elderly people.

CHAPTER 8 Man vs. Manmade

The emergence in the early twenty-first century of a new form of intelligence on Earth that can compete with, and ultimately significantly exceed, human intelligence will be a development of greater import than any of the events that have shaped human history.
—Ray Kurzweil, *The Age of Spiritual Machines*

For once, the mighty German soccer team looks outclassed on the Osaka pitch. When a Japanese defender boots the ball into the visitors' half, they are left standing flatfooted. A member of the home team is first to reach the ball and pauses briefly as he selects his spot. He rockets it toward the German goal.

Gasps escape from the spectators as the ball looks likely to just curl to the wrong side of the post. But no, the ball bounces off a hapless German defender's feet and straight into the net. The home crowd goes crazy. But amidst the fist pumping, camera flashes and snapping shutters, the players emit not a flicker of emotion.

No matter, RoboCup is won. Japan is the world champion.

The nineteen-inch-tall Maradona is Vision Nexta, a humanoid robot with a transparent dome head, flashing diamond eyes, broad shoulders and mighty arms over a narrow waist. For the second year in a row, this Team Osaka robot has won the humanoid contest at RoboCup, which is a sports tournament, engineering and AI challenge, academic symposium and trade show all rolled into one.

It is far from being an underground geekfest in a hobbyist's garage. RoboCup is the world's most popular autonomous robot competition. Teams of robots are pitted against each other while researchers compare notes on making ever-more agile and intelligent mechatronic athletes. The Osaka RoboCup, the ninth since

Humanoid soccer-playing robot Vision Nexta is one of the biggest stars of RoboCup, an international autonomous robot sport tournament. Developed by academia-industry consortium Team Osaka, Vision Nexta has an omnidirectional head camera and dynamic moves like diving to block balls from entering the net. At RoboCup 2005 Osaka, it beat all other contenders in its league and took home the Louis Vuitton Humanoid Cup (opposite).

its launch in 1997, drew 333 teams of robot builders from thirty-one countries around the globe.

A staggering one hundred and eighty thousand visitors, mostly Japanese, flocked to the Intex Osaka exhibition hall during the five days of the event, an average of thirty-six thousand a day—more than enough to fill a major sports stadium. Inside the four cavernous exhibition halls taken up by the tournament, visitors watched robot teams compete for the Louis Vuitton Humanoid Cup, a crystal glass orb nestled in a red-lined case stamped with the logo of the luxury bag maker.

Children from around the world crowded the edges of soccer pitches to watch "kid size" robots compete at penalty kicks, passing accuracy and dribbling balls around obstacles. Other visitors watched Sony Aibo robot dogs zipping around ping pong table-sized pitches with soccer balls tucked neatly under their muzzles. Though machine soccer—divided into small and mid-sized, humanoid, four-legged and simulation leagues—is the event's main draw, RoboCup has grown to include RoboCup Rescue, a contest for disaster-rescue robots clambering through debris, and RoboCup Junior. The latter is an international education initiative in which teams of schoolchildren try to produce the best soccer-playing, rescuing and dancing robots. In 2005, schoolchildren from eighteen countries made the trip to Osaka from as far away as Finland and Iran.

RoboCup cofounder Minoru Asada of Osaka University believes that the event's robots will be a great source of inspiration for tomorrow's roboticists: "Just as there are many researchers who longed to become robot experts because of Mighty Atom, there will be many kids who want to make robots after seeing Vision's performance."

The Dream Team

Behind the scenes, the challenge of building artificial athletes was evident. In the cordoned-off "paddocks" behind the pitches, engineers struggled to revive dead droids, tweak computer code, debug systems or simply rest. Other participants were slumped over their laptops from jetlag and sheer exhaustion, surrounded by bottles of green tea, rolls of black duct tape and jungles of wiring. Battle-weary robots were cooling their circuits in front of portable fans.

In the popular RoboCup Four-legged League, Sony Aibo dogs that have been reprogrammed for faster movement scoot around a small pitch chasing the specially colored ball. The Aibos operate entirely without human control, and can wirelessly communicate with teammates. The detailed rulebook includes provisions for penalty kicks and tiebreaking shootouts.

Strain was even visible on the faces of Team Osaka, the "dream team" that built the two-time humanoid soccer champ, Vision. Drawn from a consortium of small and midsize businesses and researchers in the Osaka area, the team represents the aspirations of smaller Japanese projects to build humanoid robots in the shadow of global manufacturing titans like Honda, headquartered in Tokyo.

Team Osaka leader Nobuo Yamato is also president of Osaka-based robot producer and marketer Vstone Co. A graduate of the National Defense Academy, Yamato is committed to making humanoid robots commercially profitable, though he says he places as much emphasis on the importance of giving people dreams as making money. He hastily formed Team Osaka in response to a call for a 2004 RoboCup team by the Osaka municipal government, and received ¥15 million in city backing. Winning a robot tournament was a perhaps natural goal for someone from a city that prides itself on being down to earth and the polar opposite of the capital.

"Osaka isn't really a center for IT-related work," says Yamato, who believes smaller humanoids are the best way to commercialize home robots. "This is a cultural thing. Osaka people love making things that you can see, not invisible stuff like content, which is big in Tokyo."

Aside from being the world's biggest robot sport tournament, RoboCup is an education and research initiative. Launched in 1998, RoboCup Junior is designed to get primary and secondary schoolchildren interested in robotics and RoboCup. Kids can also take part in dance and rescue robotics competitions.

Other key team members include Hiroshi Ishiguro, a robotics professor at Osaka University who developed the omni-directional head camera that helped win the cup; in fact, a plan to commercialize his sensors led to the foundation of Vstone as a business-academia collaboration in 2000. Three years later, they produced Robovie-M, an autonomous, bipedal and acrobatic little humanoid that could pick up mini soccer balls and toss them overhead. It had an eleven-inch aluminum frame and looked like one of the fighting robot kits that are now very popular in Japan, like Kondo Kagaku's KHR-1.

But it took Tomotaka Takahashi, a "robot creator" and head of Kyoto University-affiliated startup firm Robo Garage, to turn Vstone's little footballer into the work of art of plastic and carbon fiber that is Vision. With his movie star looks and finely trimmed goatee, the thirty-year-old Takahashi is probably the hippest robot researcher in Japan. As a Kyoto University student inspired by anime about Mighty Atom as well as giant robots like Iron-man No. 28, in six months he designed and built a ten-inch tall radio-controlled humanoid robot, Magdan. That and several other striking robot creations have earned Takahashi profiles in *Time* and *Popular Science*. He was responsible for Vision's anime-esque appearance.

A member of Team Minho from Portugal's University of Minho makes last-minute adjustments before competition begins in the Middle Sized League, which pits teams of four robots against each other. Since they travel on wheels, the action is far quicker than in the Humanoid League.

"My main purpose in building robots—to make them look more friendly, sophisticated and natural—is different from other researchers," says Takahashi, who still makes bipeds in his bedroom by molding their carbon shells with a vacuum cleaner and a portable stove. "That's one of the most important parts about home-use robots one day becoming members of the family. If it's too ugly or mechanical-looking, people can't feel like it's a friend. The mental aspect is the biggest obstacle when we talk about the relationship between robots and humans."

Team Osaka's first soccer playing Vision robot was fifteen inches tall and weighed a little more than five pounds. Its 40 MHz CPU controlled a body with twenty-three degrees of freedom, allowing it to stand up from a prone or supine position. Its body was free of visible wires, gears or other mechanisms. Its head camera and object-detection algorithms based on color differences allowed it to locate and approach the ball, line itself up for a kick, during which its eyes changed from red to blue, and boot the ball into the net. One endearing characteristic was the way it would stop in front of the ball and bend over slightly to see it.

That first Vision won five events at the RoboCup Japan Open and took the humanoid championship at the 2004 RoboCup in Portugal on its first try, beating out eleven other teams. In 2005, its successor Vision Nexta was bigger and heavier but with eight times the processing power and four times the camera resolution.

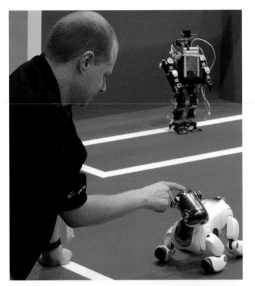

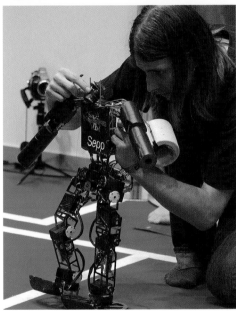

Although the human participants are not allowed to intervene while their robot teammates are dribbling soccer balls on the pitch, they are always waiting on the sidelines to fix buggy droids.

At a press event prior to the 2005 RoboCup, Vision, defending the goal, looked on almost dejectedly as Nexta nonchalantly put the ball in the back of the net.

Vision Nexta can walk over uneven terrain and introduce itself with a new voice function. It also has goal-keeping skills to match its footwork, diving left or right to stop balls like a human player. Of course, the cute factor has not been overlooked. Having introduced itself, Nexta makes a little bow with one arm tucked behind its back, like a maitre d' with fluorescent azure eyes.

In the two-on-two humanoid competition, newly introduced to RoboCup in 2005 as the event closest to a human soccer game, Vision Nexta proved a pint-sized Pele of the pitch, dominating games. Team member Ishiguro, however, thinks that the Vision Nexta team's most important achievement was not winning the Louis Vuitton crystal but bringing mass production of humanoids closer to realization. "It didn't matter to me that we won the RoboCup championship," says the engineering professor. "The real importance of Vision Nexta is that we programmed it using an average PC with the Windows XP operating system. Why do we use IBM PCs and Windows? Because everyone can use them. This is very similar to the beginning of personal computers."

But just why are robots dribbling soccer balls up the mini pitches of RoboCup? What's the point of pitting teams of robots against each other?

RoboCup was founded as a challenge to the world's scientists and engineers to produce a team of machines that can take on and defeat the best human soccer players in the world in less than fifty years. Its founders have a vision of eleven humanoid robots entering a stadium, spreading out on a soccer pitch and going head to head—or mind against processor, one might say—against the reigning World Cup championship team of 2050. And, with a crowd-pleasing display of robot passing, feinting, shooting and defending fully in compliance with FIFA rules, beating them.

But what would it really mean if robots *could* beat humans at soccer? Would people eventually shrug it off as an everyday fact of life, like cars traveling faster than humans, or robot assembly lines welding vehicles together faster than factory workers? Or would it rank with another particularly historic human defeat?

Some of the humanoids participating in Robo-Cup Osaka 2005 pose for photographers. The goal of RoboCup is a bold AI and engineering challenge—to produce a team of soccer-playing robots that can take on and defeat a World Cup champion team of humans by 2050.

From Pawns to Penalty Kicks

In 1997, world chess grandmaster Garry Kasparov, one of the greatest players the game has ever seen, was beaten by an IBM supercomputer called Deep Blue. For forty years chess had been the classic proving ground for artificial intelligence research. Computer scientist Claude Shannon wrote of the value of creating chess-playing systems in 1950: "The problem is sharply defined. It is neither so simple as to be trivial or too difficult for satisfactory solution. And such a machine could be pitted against a human opponent, giving a clear measure of the machine's ability in this kind of reasoning."

Computer chess is known as a "standard problem"—a clearly defined, realistic challenge, often with an unambiguous win/lose outcome, which AI researchers can work on together. British mathematician Alan Turing, the father of modern computer science, first conceived of chess-playing computers in the mid-1940s and wrote the first computer chess program in 1950.

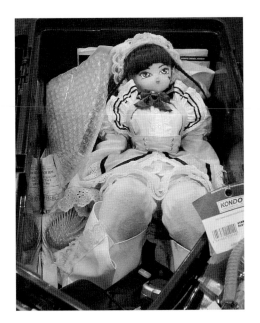

RoboCup sometimes hosts other robot sport tournaments like Robo Fight, a remote-controlled robot combat contest featuring elaborate fighting techniques and poses. One popular gladiator is Motohiro Nakamura's doll-like Cutie Bunny, seen here in its carrying case cum toolkit.

That year also saw the publication of Turing's seminal paper on AI in the journal *Mind*, "Computing Machinery and Intelligence," which began with the question "Can machines think?" In response, it introduced his famous Turing Test, an examination of a computer's ability to mimic human conversation. A human engages in a conversation with two unseen correspondents, one a machine and the other a person, by means of a text interface. If he or she cannot judge which is which, the machine has passed the test.

One of the first modern conversational programs, or "chatterbots," was ELIZA, developed in 1966 as a virtual therapist by MIT computer science professor Joseph Weizenbaum and named after the uneducated Eliza Doolittle in George Bernard Shaw's play "Pygmalion." It employed simple string substitution and rote responses based on keywords, but conversations could descend into nonsense. An example of a similar application:

> Hello, I am Eliza.

* Hello.

> How are you today. What would you like to discuss?

* The weather.

> Tell me more . . .

* It's really been awful lately.

> That is quite interesting.

* Do you think so?

> Oh . . . think so?

* Yes.

> I see.

* What?

> Have you asked anyone else?

* What do you mean?

> We were discussing you, not me.

* Whatever!

> OK . . . "What do 'me' mean." Tell me more.

Oddly enough, ELIZA actually fooled some users in the 1960s.

Turing predicted that computers would be able to pass the test with one-third of human judges by 2000, but this proved overly optimistic. Though futurist Ray Kurzweil has predicted real conversational computers by 2020 based on the exponential growth in microchip computing power, for now Turing machines remain an unsolved standard problem of AI.

Computer chess, however, is a different story. The Kasparov-Deep Blue games, heralded by *Newsweek* magazine as "the brain's last stand," were the first time a computer had beaten a reigning chess grandmaster under normal time controls. Kasparov, stunned that the computer failed to fall for a trap that would usually snare artificial players, controversially suggested IBM had concealed human grandmasters in Deep Blue's control room to intervene at key moments in the match. But he also saw something like intelligence in the machine's moves. His assistant Frederick Friedel commented: "It's a weird form of intelligence, the beginning of intelligence. But you can feel it. You can smell it."

IBM dismantled Deep Blue after it beat Kasparov (despite the Russian's demand for a rematch) and subsequent grand master-supercomputer contests have tended to end in draws. But the epic battle underscored an important human frailty that may be key if robot footballers are to trounce human opponents in a soccer game. "When Kasparov was beaten by Deep Blue, he mentioned that humans get tired, make mistakes and are prone to psychological stress, but computers don't," says RoboCup cofounder Hiroaki Kitano. "He said that was the decisive factor. I

Inspired by Japanese manga and anime, Chroino (above), Magdan (below left) and Robofie are RC robots created by Robo Garage's Tomotaka Takahashi, who also designed Team Osaka's Vision Nexta.

Tomotaka Takahashi

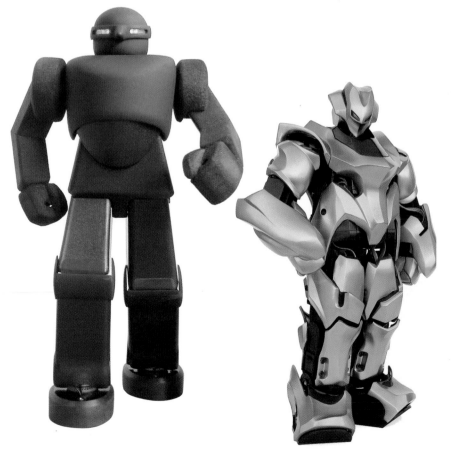

Built of aluminum, carbon and plastic, Takahashi's cute Neon evokes his hero Mighty Atom. Its rounded contours and subdued colors are designed to enhance its appeal. Manoi (far right) will be the latest Takahashi RC robot to go on sale. Marketed by Kyosho Co. as an athlete robot, Manoi features high-speed, powerful actuators and resin brackets for smooth movements.

RoboCup attracts participants from around the world. This imposing goaltender is from Iran's Sharif University of Technology.

think that could be true in a human-robot soccer game."

While a chess-playing computer program that defeats a grand-master may appear to demonstrate superior intelligence, it simply employs the brute force strategy of examining as many moves as possible.

This number-crunching is good for very specific tasks like weighing the likelihood of a possible checkmate ten moves away, but falls far short of the complexity of human thinking. Otherwise, computers would be equally good at playing other games; to take one example, computers have basically failed at the ancient Chinese board game of Go because of its extremely high number of possible moves. And, as MIT AI lab cofounder Marvin Minsky commented, "Deep Blue might be able to win at chess, but it wouldn't know to come in from the rain."

Kitano and the other founders of RoboCup want robot soccer to be the chess problem for AI in the twenty-first century. They want to take the field from abstract logic puzzles to dynamic real-world challenges. In fact, boredom was a prime factor in the conception of a tournament of machine athletes.

The Challenge

Kitano, director of Sony Computer Science Laboratories, Inc., researches what he terms "systems biology"—looking at organisms as systems from the molecular level up in order to understand the fundamental principles of life. His Kitano Symbiotic Systems Project, headquartered on Tokyo's trendy Omotesando boulevard, turns out papers on topics like the suppression of malignant cancer cell phenotypes—far from the playful realm of sporting robots. But in 1993, Kitano attended a robot competition organized by the American Association for Artificial Intelligence (AAAI), at which robots in one event faced the underwhelming task of taking a paper cup off a desk.

The machine contestants imaged their environment with video cameras then sat motionless for long minutes while analyzing the data. They would then move a few inches and repeat the process. As Asada noted, it was "not really a spectator sport, nothing to capture the viewer's imagination." Kitano, Asada and a third researcher, Yasuo Kuniyoshi, decided they could do better.

They considered several sports. The rules of baseball, it was decided, were too complex and most of the action takes place

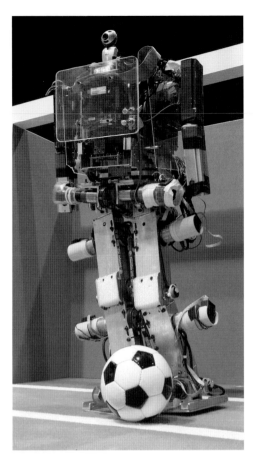

Isaac, built by students at Italy's Politecnico di Torino, tends goal. At thirty-three inches, it is one of the taller robots in RoboCup's Humanoid League.

between batter and pitcher. The act of dribbling a basketball was deemed too mechanically difficult. The game of soccer, on the other hand, with eleven players to a side trying to put a ball in the other side's net, represents a relatively simple set of parameters. It also entailed a host of important, complex and real-world robotics challenges such as collaboration, autonomous agents and real-time reasoning. Most appealing was its universal appeal—after all, over 240 million play the game regularly in more than two hundred countries. Surely this was a shared challenge researchers around the world could tackle.

In June 1993, Kitano, Asada and Kuniyoshi conceived of a robot tournament called the "Robot J-League," after the name of the Japanese professional soccer league, which had just been launched with massive publicity. But researchers in other countries responded so enthusiastically, urging that it be made an international project, that the endeavor was rechristened RoboCup, short for "Robot World Cup."

In 1997, the year Deep Blue took down Kasparov, Kitano, Asada and Kuniyoshi held the first RoboCup tournament and conference in Nagoya. Over forty teams participated before five thousand spectators. But the contest instantly caught the imagination of robot researchers and hobbyists all over the world. Today, variations of RoboCup soccer and simulated soccer games are held in venues as far flung as Atlanta, Bratislava and Isfahan. Since many researchers view computer chess as largely "solved" by Deep Blue, RoboCup has come to represent a new standard problem in AI, as well as robot engineering.

The Body Electric

The artificial players have developed markedly in their abilities since that first tournament. From trouble-prone boxes on wheels, they have become agile little bipeds that keep spectators riveted with kicks, tumbles, saves and goals—progress that owes much to attitudes instilled by RoboCup's founders.

Collaboration and sharing are crucial. Participants submit "team description papers" outlining their innovations, mechanisms, algorithms and strategies and are encouraged to produce scientific papers, which are judged for science and engineering prizes. Depending on the league, computer code can be shared entirely, something especially important for the simulated soccer competition, which consists of two- or three-dimensional games played by software agents, projected on large screens. As University of New South Wales professor Claude Sammut noted at the 2004 games: "Our first time in the competition was in 1999—and the robots were hopeless. Since then, the robots themselves haven't changed that much, but the performance is dramatically better because we've learned how to write the software in a much more clever way."

RoboCup co-founder Asada is a thin, wispy-haired 53-year-old,

Team Osaka's Vision Nexta robot does some fancy footwork to outmaneuver the University of Freiburg's Team NimbRo players and clinch the championship in the 2005 RoboCup Humanoid League, its second tournament victory in a row.

head of his own laboratory at Osaka University's Graduate School of Engineering. Sipping a cup of green tea, he reflects on how much robot soccer players have evolved since the first RoboCup in 1997. Reporters at the event, bored by the buggy, frozen robots and lack of action, were prone to ask, "Has the game started yet?" Some attendees were disappointed by the level of technology, but others were stimulated to build better players. For Asada, the long road to the real World Cup is itself the aim of the entire exercise: "The purpose of RoboCup is not the final performance, but the process we're going through to share the dream," he says.

Yet Asada is certain RoboCup will meet its ambitious challenge of pitting machine against man on the soccer pitch—and he believes the robots will have an even chance of winning. "The final goal is to build a team of eleven humanoids that can beat a human World Cup champion team," he reiterates with a warm, broad-faced grin. "This is like NASA's projects to send a man to the moon or to Mars."

For his part, Asada's research centers on machine intelligence, especially "emergent intelligence" that is grounded in the body. Asada believes intelligence emerges *because* of the body. He is driven by a fundamental curiosity about what we are as a species and sees robots as the best scientific instrument to answer that question: "Personally I want to know what we are—what human beings are. Building robots brings us closer to knowing." Stu-

dents in his lab are trying to teach machines to acquire human language just as infants do, starting with rudimentary syllable learning and artificial throats.

How do we make a robot that can really understand what an apple is, Asada asks. "Only from having a body—that holds the apple, touches it, smells it and bites into it—do we finally learn what an apple is. I believe that the semantics of recognition come from our corporeal experience, and not from the symbolic confines of a computer's interior."

RoboCup players will also need bodies to match their task. Asada suggests equipping robot athletes with artificial skin to make physical contact with human opponents or teammates as painless as possible; one spin-off could be commercializing the skin for robots serving in hospitals. But for robots to take on a World Cup powerhouse like Brazil, the game would also have to be a fair match, and though robots might not feel pain, they would have to be programmed to somehow suffer, for example, from a collision with their human opponents. Asada cites the self-preservation principle in Isaac Asimov's Third Law of Robotics: "A robot must protect its own existence, as long as such protection does not conflict with the First or Second Law [never harming and always obeying humans]."

Consciousness will be an absolute necessity for robots to play a convincing game of soccer against human rivals. Feinting with the ball, for instance, requires a psychological understanding of an opponent's immediate intentions from reading body language and the ability to deceive based on that understanding. "You need to have a sense of self to fool others. You also have to be able to stand in their shoes, see what they're seeing, know who they are. Whether this is something which is genetically or experientially stipulated is an essential point."

Besides a healthy obsession with the World Cup challenge, Asada may have another motive for pursuing his research. Like many Japanese of his generation, as a child he was enthralled by Osamu Tezuka's Mighty Atom, and eagerly read the manga books and watched the animated series on TV. He suggests that most Japanese robotics researchers "are working toward an ideal of robotics that is at some level influenced by Mighty Atom," which explains why their creations "may be a bit more vibrant, and capable of meaningful communication with human beings than their Western counterparts."

Asada sees two other factors behind robot development in Japan. One is that, in contrast to Western perceptions of technology and artificial life as hostile or evil, the cultural and religious predisposition of Japanese is toward animism and the attribution of souls or minds to inanimate objects: "Japanese people think everything has a mind, everything has a soul. This glass, or the walls, everywhere," he says. "People think: 'This is not a machine, but a friend or partner.'"

The other factor is the art of *monozukuri*, what Asada describes as a national love of making things, which he believes is stronger than in Europe or America, and is reflected in Japanese fine art. He also thinks everyday robots will be accepted much more readily in Japan than elsewhere. "I believe this will be the first nation where we will use personal robots in our daily lives," he says. "So therefore it's the role of Japan as the Robot Kingdom to disseminate to other nations all the information that is learned from daily life with robots, good or bad. That is our obligation."

CHAPTER 9 Android Dawn

Like all things, advanced androids will have a Buddha nature—the potential for enlightenment. It exists everywhere, not just in robots that look like humans. It is in industrial robots. It is in hunks of steel. It is in pebbles. The Buddha nature is in everything.

—Masahiro Mori

To play soccer against human athletes and to work in human environments, robots need humanlike bodies. To be better accepted by people, they need humanlike faces. Extending this line of reasoning by roboticists, we arrive at the somewhat disquieting conclusion that the ideal robot would look exactly like a human. In other words, an android.

Androids are robots that look just like real people. They are much closer to the ideal human form than traditional robots. The term, from the late Greek *androeides*, means "manlike." It also encompasses the notion of synthetic beings that are mainly organic instead of mechanical.

Androids first appeared in science fiction nearly forty years before the android-like robots of Čapek's *R.U.R.* The name of the French symbolist writer who introduced them was a regal mouthful: Count Jean Marie Mathias Philippe Auguste de Villiers de L'Isle-Adam. His 1886 novel *L'Eve future* (*Tomorrow's Eve*) is as longwinded as his name, but is basically about a fictionalized Thomas Alva Edison's creation of a humanlike, electro-magnetic robot woman. Edison produces the robot for an English aristocrat upset that his fiancée lacks brains to match her beauty. The inventor gives him the perfect woman, a beautiful and intelligent android named Hadaly, boasting, "Her heart never changes; she hasn't got one."

Hadaly, a male fantasy of engineered female perfection, and a nineteenth-century retelling of the Pygmalion myth, heralded countless other androids in fiction. Some were male. One of the most famous in the West in recent years was Data, the pale-skinned officer with a positronic brain aboard the starship Enterprise in the *Star Trek* television franchise. In Japan, "android" has been primarily translated as *jinzo ningen*, or "artificial human," the early term for robots that was inspired by *R.U.R.* Although he was not described as such, Mighty Atom, made to look just like a human boy, is really an android.

Tezuka's famous student Shotaro Ishinomori penned many manga featuring androids and cyborgs including the popular *Jinzo Ningen Kikaida* (Android Kikaider), about a guitar-strumming android hero named Jiro who has "a human soul in a manmade body." Published in the early 1970s as Kato's WABOT was nearing completion, the comic anticipated the coming of human-like robots. "Unfortunately, androids are still at the experimental stage," Ishinomori wrote to his boy readers in an introduction citing Čapek, "but it's only a matter of time until they are put to practical use."

Twenty-first Century Eve

Unlike robots, few androids have stepped from science fiction into the real world. But—in Japan, at least—that is changing. At the 2005 Aichi Expo, humanoid robots from labs throughout the country were put on display to the delight of millions of visitors. The exhibits came in all shapes and sizes: they moved on wheels, walked on two legs, looked like lovable little dolls or fantastic mechanical warriors. All, however, were instantly recognizable as artificial organisms.

Except one. This one had moist lips, glossy hair and vivid eyes blinking slowly as it gazed around the room. Seated on a stool with hands folded primly on its lap, it was dressed in a bright pink blazer and grey slacks. True, there was something not quite right about the gaze, the mouth and the skin. But for a mesmerizing instant from fifteen feet away, the senses were deceived and the Repliee Q1expo was virtually indistinguishable from a typical Japanese woman in her thirties.

Robotics laboratories around the world have turned out numer-

Which is which? NHK news announcer Ayako Fujii (left) poses with her robot clone, the android Repliee Q1expo, created for the Aichi Expo outside Nagoya. A modified version called Repliee Q2 (opposite) is the most lifelike humanoid machine in the world.

ous artificial hands, heads and legs, but the results have been mixed: some feature very humanlike and elegant abilities in grasping, expression or locomotion, and others are slow and jerky. Few, however, are serious attempts to *look like* their human counterparts. The complexities of basic motion control in robots seem to have kept developers far too busy to bother with verisimilitude.

Repliee, however, is perhaps the closest thing in the world to a robot that can pass for a human. Unlike the sophisticated but non-interactive animatronic attractions at Disney theme parks, for example, the robot is aware of its surroundings and can respond to stimuli. It talks, gestures and interacts. It watches, it waves, it shifts in its seat, it acts like a real woman. Repliee is a twenty-first century Hadaly.

Android Science

Repliee's creator, and the director of Osaka University's Intelligent Robotics Laboratory, Hiroshi Ishiguro has a high furrowed brow beneath a shock of inky hair and riveting eyes that seem on the verge of emitting laser beams. When posing for photos, he assumes a practiced gaze of intensity, even anger, under knotted eyebrows, as if reveling in the role of an eccentric scientist. But why is Ishiguro intent on producing robots that are as close as possible to human beings? His answer is simple: "android science."

The argument for making robots anthropomorphic tends to focus on building machines that move like us so that they can operate in environments designed to suit the human body. For domestic robots, the classic example is the staircase—removing this convenient architectural feature to accommodate wheeled or other robots is impractical. Thus, bipedal is best.

Another widely held view is that the more naturally and smoothly robots can interact with people, the more effective they will be as tools, entertainers, helpers or companions. While most research on such interaction has concentrated on the behavioral aspects of robot development—motion control, degrees of freedom and artificial intelligence—Ishiguro believes that robot *appearance* is just as crucial.

"Appearance is very important to have better interpersonal relationships with a robot," says Ishiguro. "Robots are information media, especially humanoid robots. Their main role in our future is to interact naturally with people."

Ishiguro's approach has much in common with an argument for behavior-based AI systems put forth in the 1990s by Rodney Brooks of the Massachusetts Institute of Technology. Brooks rejected designs for machines that could *think* intelligently in favor of ones that could *act* intelligently. His view of human cognition as an activity too poorly understood to emulate scientifically struck a chord with many researchers who had to admit that decades of attempts to replicate human-level intelligence in machines had failed.

Smart robots, the Brooks school held, should be capable of meaningful action and interaction in a real environment, and not some closed system of abstract logic and symbols. This behavioral approach to AI has been very successful, spawning intelligent systems in fields as disparate as space exploration and healthcare.

Ishiguro has taken Brooks's emphasis on artificial systems being *embodied* (a physical body in the world) and *situated* (sensory input influencing behavior) and added an extra dimension to the equation—verisimilitude. But his quest to build very lifelike robots has been shaped by his own experience.

Although Ishiguro grew up building his share of Gundam models, he was more focused on philosophical questions about life than inventing robots. Mild colorblindness forced him to abandon his aspirations for an art career, however, and he found himself drawn to computer and robot vision, building a guide robot for the blind as an undergrad at the University of Yamanashi; Mitsubishi's Wakamaru is partly based on a humanoid he later developed called Robovie. After traveling this path, Ishiguro now sees robots as the ideal vehicle to learn more about ourselves and questions about human nature.

An avid fan of Star Trek's Data, he also believes that only very humanlike machines, i.e., androids, can elicit natural responses in people, and can thus integrate into human society. But building a lifelike mechanical person, a common but distant goal in robotics, presents an extremely complex and perhaps impossible challenge.

Ishiguro focuses on fundamentals of interaction. He sees human appearance and behavior as key in creating androids. To successfully emulate these, he yokes the disciplines of robotics and cognitive science. The construction of a robot based on knowledge of human behavior from cognitive science can produce a more humanlike robot. In turn, cognitive science research can use that

◀ Though it is confined to a stool, Repliee Q2 is a fully interactive android. It is aware of people around it through an exterior sensor array that includes cameras and microphones. Repliee's main purpose, though, is to serve as a tool in robotics and cognitive science research.

robot as a test bed to verify theories about human perception, communication and other faculties.

This interdependent, cross-fertilizing and novel approach is what Ishiguro describes as android science.

Girl, Replicated

When Ishiguro makes an android, one key factor in his approach is that he models them on real people. He describes the process of android construction as "making a copy of an existing person." For Repliee Q1expo, Ishiguro and his students chose popular Japanese TV presenter Ayako Fujii of NHK's *Weekend Kansai* show. Fujii agreed to let codeveloper Kokoro Co. fashion a specially pigmented silicone skin based on her own. In effect, she allowed herself to be cloned in robot form.

The process employs shape memory silicone molds, which are applied to various areas of the human model's body. The molds are then used to create plaster body parts that fit together to form a complete dummy of the model. A metal and polyurethane body is developed from this, and a soft silicone skin a fifth of an inch thick is the last to be added, covering the upper torso, face, neck, forearms and hands.

To achieve smooth upper-body movement in Repliee, Ishiguro equipped it with forty-two small, quiet air servo actuators. Four are devoted to the mouth for the important challenge of synchronizing speech and lip movement. Since the actuators are powered by a fridge-sized external air compressor, locomotion was sacrificed. Similarly, most of the android's control elements and sensors are offboard. It runs on eight PCs and is surrounded by an array of floor sensors to track human movement, video cameras to detect faces and gestures and microphones to pick up speech. A cutting-edge combination of robotics and embedding computing devices into the environment, Repliee has none of the bulky machinery of most humanoids, but a high degree of motion range at thirty-one degrees of freedom.

After producing their realistic-looking android, the team concentrated on animating it. They used a motion-editing system and are also working on three-dimensional motion capture, which involves placing electronic markers at key places on a human performer and having a computer track and reproduce the movements in the robot. An important element in making the android

"Two hundred years ago, art and technology ▶ were integrated. We didn't distinguish them," Repliee's creator Hiroshi Ishiguro, a professor in Osaka University's Graduate School of Engineering, says of making androids. "I think it's time to merge once more. Therefore, I'm very interested in android science, which integrates engineering and science and maybe art too."

Actroid, an android receptionist developed by Kokoro Co. and Advanced Media Inc., provided directions at the Aichi Expo. Its voice recognition engine allows it to understand and respond in English, Japanese, Chinese and Korean without prior training, while exhibiting humanlike gestures and emotions. Kokoro's Tatsuo Matsuzaki admits: "The greatest words of praise would be 'I thought she was a real person.'"

appear as humanlike as possible is that Repliee draws upon a library of these "pre-learned" gestures—such as the involuntary shoulder movements we make when breathing. The fidgety, slightly uncomfortable behavior is wonderfully human.

The finished product, complete with hair, makeup and clothing, as well as lifelike expressions, really does look like the real Ayako Fujii. Though it remains confined to a stool, and is unable to move its legs, its distributed sensors allow it to learn about its environment and interact with humans. Repliee recognizes when people point in a certain direction or bow to it, an action that occurs with such frequency that its creators are no longer surprised.

"I was developed for the purpose of research into natural human-robot communication," says Repliee, raising its arm in an instantaneous response to a touch picked up by skin sensors.

If seeing the android alone is startling, seeing it together with Fujii, its human "parent," is downright eerie. The sensation is compounded by its name. Repliee was taken from the French *repliquer*, to replicate, but evokes the "replicant" biological androids of Ridley Scott's 1982 science fiction masterwork *Blade Runner*. In the film, rogue robots in a dark, dystopic Los Angeles of 2019 are so indistinguishable from people—the slogan of their manufacturer, the Tyrell Corporation, is "more human than human"—that elaborate "empathy tests" must be administered to tell machine from man. In a world of perfect simulation, the film asks, who is real?

Enter the Uncanny Valley

Blade Runner explored a problem that roboticists have also addressed: Can an android be too human? According to some, the more robots resemble humans, the more their subtle imperfections make us feel uncomfortable, engendering a deeply negative response in us. This phenomenon is known as the Uncanny Valley.

The concept was suggested in a 1970 paper by Masahiro Mori, a contemporary of Ichiro Kato, who was motivated by an interest in human anatomy and robotics. Mori grew up in heavily industrialized Nagoya, where an early fascination with radios and electricity found him working part-time as a flutist in the local symphony so he could buy electronics texts. As a student in Tokyo, he operated machinery in chemical and food-processing plants to understand automated control. He concluded that to create better forms of automation, machines must have hands and fingers, just like humans. Mori, a devout Buddhist, wrote a provocative book while he was a robotics professor at the Tokyo Institute of Technology entitled *The Buddha in the Robot* that goes a step further, arguing that machines can have souls as well.

During his childhood in the late 1930s, Mori visited a waxworks display and was struck by the eeriness of the human figures. This experience later inspired him to hypothesize the Uncanny Valley as a conceptual graph in a paper entitled "*Bukimi no tani*"—literally, the "Valley of Eeriness." He says that "as robots appear more humanlike, our sense of their familiarity increases until we come to a valley."

He compared the familiarity, or positive feelings we have when seeing humanlike entities, to climbing a mountain. Even though the distance from the summit—the analogy to a real human—decreases, there is a valley encountered en route. The valley appears when a robot's charmingly human traits are overwhelmed in the mind of the viewer by the degree to which it falls short of being human. In other words, when it looks too real, yet clearly isn't, people's emotional empathy ceases.

Mori used realistic prosthetic hands as an illustration. With

Clad in a miniskirt and jacket, this emcee Actroid is a sexier version of its information booth counterpart. Though it can speak with a prerecorded female voice and wave like a human, it has no sensors. But crowds at exhibitions where it has presented other robots on stage are more struck by its glistening eyes and other lifelike features.

lifelike fingernails, veins and skin pigmentation, they may even be indistinguishable from a real hand at first glance. But shaking such an appendage when greeting someone is a shock: we are surprised by the hardness and lack of warmth. This is the Uncanny Valley.

In his graph, Mori situated stuffed animals and toy robots on the first incline, and bunraku theater puppets on the steeply ascending slope after the valley. He also argued that movement amplifies the uncanny effect. A second graph illustrated this, with completely non-anthropomorphic industrial robots at the foot of the mountain, humanoid robots on the first slope, sick people on the other side of the valley and healthy people at the peak. Near the bottom of the valley were moving prosthetic hands.

Taking this even further, Mori reasoned that since movement tends to be a sign of life, the walking dead would be the most unnatural and thus repulsive human-like creatures—and populated the lowest depths of the ravine with zombies. He saw, then, the Uncanny Valley as the landscape of death, and the act of dying as falling off the summit into the abyss.

Mori concluded that the natural human aversion to death explains the mystery of the Uncanny Valley phenomenon, and speculated that it may also be important for our self-preservation. He warned roboticists that building a completely humanlike robot would simply reinforce the eeriness of the valley, and exhorted them to set their sights on the area before the valley. His thinking is that people will always feel closer to a robot that is more like a machine than a person.

"In 1970, humanoid robots existed in manga and anime but not in real life," says Mori, who at nearly eighty years old is still avidly investigating spirituality and machines at his personal Mukta Research Institute in Tokyo. "There was no reaction to the paper back then. But now there's worldwide interest in the idea because humanoid robots have been realized. Another reason is that while its mechanisms may be inexplicable, the Uncanny Valley seems to hide something deeply connected to human life or human psychology."

Disaster and Redemption in the Valley

Some researchers have dismissed Mori's theory as pseudoscientific, others see uncanniness as simply a question of aesthetics.

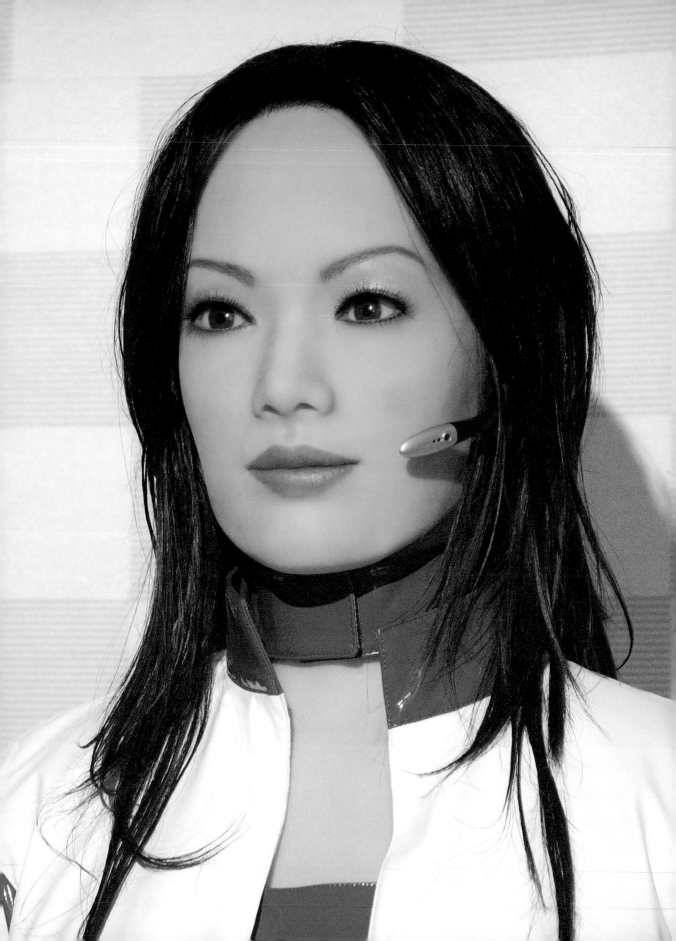

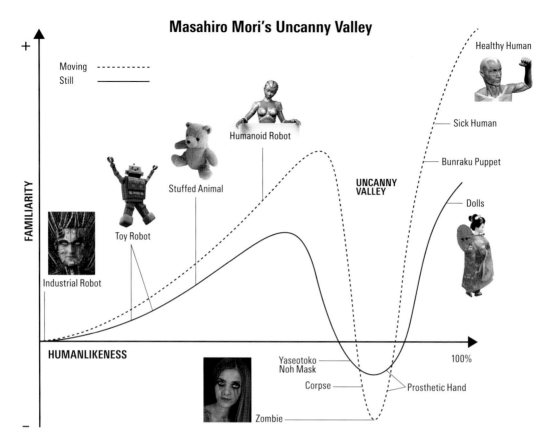

Masahiro Mori's Uncanny Valley

Moving ---------
Still ——————

FAMILIARITY

+

−

Healthy Human

Sick Human

Bunraku Puppet

UNCANNY VALLEY

Dolls

Humanoid Robot

Stuffed Animal

Toy Robot

Industrial Robot

HUMANLIKENESS

100%

Yaseotoko Noh Mask

Corpse

Prosthetic Hand

Zombie

First published in 1970 in the journal *Energy*, Masahiro Mori's conceptual graph of the "Uncanny Valley" phenomenon compares increasing human likeness in robots and other anthropomorphic objects with positive responses in people. Familiarity increases until subtle differences in appearance compromise our comfort zone and the robot seems eerie; movement amplifies the effect. Today the Uncanny Valley is a commonly known design consideration for humanoid robots and artificial characters.

Yet there is anecdotal evidence of disastrous results when the phenomenon is ignored.

The Japanese animated film *Final Fantasy: The Spirits Within*, based on one of the most popular video games of all time, managed to bomb at the U.S. box office because, as some argued, it slipped into the Uncanny Valley. Though director Hironobu Sakaguchi and some 130 visual effects staff tried to create the first photorealistic humans in a motion picture from computer-generated imagery (CGI), viewers reacted against the stiff, mannequinlike characters, their blank eyes and stony skin. One of the costliest failures in U.S. film history, *Final Fantasy* helped bankrupt producer Square Pictures, Inc.

One of Repliee creator Ishiguro's early androids was cast from his then four-year-old daughter. Equipped with few actuators, its dull facial expressions and jerky movements proved so uncanny that the girl later refused to go to her father's lab because her scary robot doppleganger was lurking there.

"If a robot is very robot-like, we never apply our human model to recognize it," Ishiguro says. "But if the robot looks like a human, we apply a human model to recognize that robot. Therefore, we notice small differences between an android and human."

How can roboticists help their creations avoid the Uncanny Valley? While Mori focused on how movement in humanlike prostheses and machines can emphasize their uncanniness, Ishiguro has found that motion itself can help a lifelike android pass for human.

In a simple android science experiment that he called a "Total Turing Test" after Alan Turing's famous AI evaluation proposal, he had twenty subjects each sit in front of a curtain. They were told that the curtain would be pulled back for two seconds—during which they were to identify the color of a cloth hung behind it. Unknown to participants, the android Repliee was also sitting behind the curtain, either motionless or exhibiting the kind of "micro movements" that people unconsciously make—such as shoulder movements when breathing. When the android was static, 70 percent of the subjects realized within the two seconds that they were looking at a robot. But when Repliee would turn its head or otherwise move slightly, 70 percent of the subjects did *not* realize it was an android.

Ishiguro believes that the experiment explains two things about the Uncanny Valley: first, that people naturally tend to see humanlike androids as human beings, but subtle differences make them uncanny; and second, that people expect a natural correlation between appearance and behavior– they want robots to behave as they would expect mechanical beings to behave and feel uncomfortable if they don't. Simply put, this means when both humanlike appearance and humanlike behavior are well matched, robots can start to escape the confines of the valley.

Ishiguro has extended Mori's graph of the valley by combining these two hypotheses, adding a third dimension to the chart. While the horizontal axis indicates similarity of appearance and the vertical axis indicates familiarity, Ishiguro's added axis of depth graphs similarity of behavior. This 3D conceptual graph, the "extended uncanny valley," may prove key in ensuring robots steer clear of the ravine of death. Instead of a sharply falling slope in the middle of the curve, it has an upward-sloping "synergy hill" that is the product of the synergy effect.

Ishiguro and his colleagues have also performed gaze direction experiments suggesting people can unconsciously react to androids on a social level even if they are consciously aware they

Tokyo Institute of Technology emeritus professor Masahiro Mori, a devout Buddhist, believes robots have the potential to achieve enlightenment. He has recently proposed that instead of humans, the faces of Buddha statues should occupy the highest point on his graph's curve, "as an artistic expression of a human ideal."

When visitors to the Tokyo University of Science walk through the front door, android receptionist Saya is the first to greet them. Created by professor Hiroshi Kobayashi's laboratory, it can express six emotions via facial expression and respond to questions or comments with about seven hundred stock phrases in Japanese. The robot even has its own nametag and hairbrush.

are dealing with a machine. One hypothesis about why a person breaks eye contact while conversing is that it is a "social signal" indicating he or she is thinking. In the experiment, a group of Japanese subjects was found to break eye contact and look down when each spoke with a researcher. But they did the same thing when conversing with an android, leading the scientists to claim that "the breaking of eye contact can be an evaluation of an android's humanlikeness."

The experiment also shows that certain body parts are crucial for a lifelike android. Since people tend to focus on a partner's eyes in conversation, Repliee's eye and eyelid mechanisms still have to be improved, and the corneas lack the proper moist appearance. For the larger issues like giving Repliee the ability to walk, Ishiguro is designing special DC motors and smaller actuators to eliminate the need for a large air compressor, but he admits this is a major engineering challenge. Yet as Repliee shows, combining androids with external sensing devices like cameras will vastly improve robots' perceptual and interactive abilities.

If Ishiguro and his successors meet their goals, androids may one day really become so lifelike that *Blade Runner*'s "more human than human" robots will live among us. But curiously,

Ishiguro dismisses such conjecture offhand. "Two seconds or ten seconds of confusion is possible, but a whole day is impossible," he declares. "We can improve the appearance, since the differences in the details—like between real human skin and silicone skin—are easy to recognize. But behavior and AI technology are quite a different matter. Perfect dialogue, for example, is impossible, and will remain so, even one hundred years from now." Still, Ishiguro wants his next android, a male, to be as authentic as possible. The model? Himself. Ishiguro thinks having a robot clone could ease his busy schedule: he could dispatch it to classes and meetings and then teleconference through it.

Android science is still in its infancy, but researchers hope it will serve as a tool to learn the dos and don'ts of advanced robot design. Meanwhile, scientists in fields like psychology and neuroscience are also intrigued by Ishiguro's ideas. "There are many people who are interested in android science but who don't have access to an android yet," says Ishiguro collaborator Karl MacDorman, who is investigating the possible links between fear of death and the Uncanny Valley. "An android is a kind of ultimate experimental apparatus and test bed. We need more of them."

As attempts to climb out of the Uncanny Valley continue, Ishiguro sees androids performing specific tasks in society in about five years. Mannequins like apparel dummies, he argues, have set the stage already. One example of a working android might be a museum receptionist giving directions to visitors. Kokoro, the firm that co-developed Repliee, set up such a talking information robot called Actroid, a commercial version of Repliee, in booths at the Aichi Expo that could converse with visitors in Japanese, English, Chinese and Korean.

Another pneumatically powered android receptionist called Saya was created by Hiroshi Kobayashi and has been doing similar work at the Tokyo University of Science for years. Though less sophisticated than Repliee, it is equipped with a CCD camera, motion sensor and microphone, can recognize some three hundred words, respond with seven hundred expressions and exhibit six "emotions" through facial expressions. "She's especially popular with little children and elderly," Kobayashi writes. "There are many repeat visitors and old men who are fans."

Ishiguro says he has no idea when androids might be common in homes, but similarly notes his own lab staff's acceptance of Repliee. Some have even developed "special feelings" for the android. "This happens with other robots too. I have one or two staff who are crazy about them."

Metaphysical Robots

While humanoid robotics requires the most advanced technology, any attempt to produce a humanlike machine must begin with knowledge of human beings, as both Mori and Ishiguro attest. In that sense, the android project is nothing if not an exploration of the human condition itself.

Seen from a distance, Repliee can easily be mistaken for a real woman. It represents the ultimate imitation in our modern culture of simulations and copies. If androids portend a future of indistinguishable human facsimiles, we may no longer know ourselves from machines.

If a moving, gesturing, talking android could pass for one of us—a living creature—it begs the question: "What is life?" Are we merely organic robots—animated masses of flesh, bone and blood? The android project and its goal of overcoming the Uncanny Valley can be seen as an attempt to make an artificial human being, one that transcends our own mortality, and as anddroids evolve, it will become ever more difficult to regard them as mere machines.

Ishiguro, the wizard of android science, puts it this way: "My question has always been, 'Why are we living, and what is human?'" With his next android, an Ishiguro made of circuitry and silicone might soon be answering his own questions.

Loving the Machine

Robots these days can think and act for themselves. We don't need people to tell us what to do.

—Mighty Atom

Yakuoji is a nondescript neighborhood in the heart of Tokyo. The barbershop that I have been patronizing there is equally ordinary—the candy cane pole outside, a waiting area with piles of manga and tabloid magazines, a few leather chairs. And it is during one snip-accompanied chat with the barber, not long after beginning research for this book, that I notice something remarkable. In every corner of the room are robots.

They are Gundam plastic models: the heroic RX-78-2, the evil-looking green Zaku and various other Mobile Suits in powerful polystyrene poses. To the shop's clients, these are perhaps barely noticeable, part of the background noise of Japan's character-saturated culture. But to their builder, thirty-five-year-old barber Toshiyuki Serita, they are a source of deep happiness.

Serita leads me up a shadowy flight of stairs behind the shop. Suddenly I find myself in an extraordinary space. Filling racks of storage shelves on a landing are dozens of Gundam models of various scales, all with exquisitely detailed armor and weaponry. Each has been carefully assembled by hand and carefully posed. It is a small museum, a hidden shrine dedicated to imaginary robots.

But this is only a fraction of his collection. Robots are scattered throughout his bedroom and other areas of the building, a home cum barbershop. Serita has been buying and building model kits since he was in elementary school. He owns every Gundam model ever produced, over five hundred all told, and has spent some ¥2.5

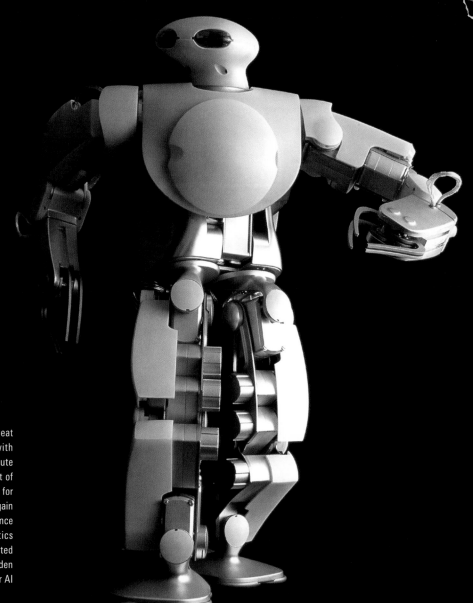

Education is crucial for Japan to repeat its success in industrial robots with humanoids. In 2006 the Chiba Institute of Technology launched a Department of Advanced Robotics to cultivate talent for new robot service industries. Students gain hands-on robot development experience with the affiliated Future Robotics Technology Center and its sophisticated Morph 3 robot, a fifteen-inch biped laden with 138 sensors that is a platform for AI and motion control research.

million in savings and thousands of hours of work on them. But he is no reclusive hobbyist. He has also traveled the length and breadth of Japan—on motorcycles he builds to his own designs.

"Just as Gundam pilots improve with battle experience, I challenge myself by building increasingly complex model robots and motorcycles. I want to see how far I can go with them," says Serita. He points to a photo of himself on one of his bikes. It was taken at Cape Soya, the northernmost tip of the Japanese archipelago.

For Serita, machines are gateways to dreams featuring faraway lands and fantastic worlds. Japan's Robot Kingdom, in fact, is founded on such shared dreams. Robot development is tied to practical needs like cutting manufacturing costs, improving productivity and finding a solution to the shrinking workforce problem, but it is dreams that drive Japan's roboticists. Scientists, engineers, government officials, and the legions of specialists who invest massive amounts of time and money on research into robots are propelled by the desire to create the imagined robot hero, friend, partner and laborer of their childhood fantasies.

How does a Kyoto electronics equipment maker show off its technical expertise? It produces a pint-sized bicycle-riding robot. One of my favorite TV ads shows a young boy who has fallen off his bike while learning to ride, his father's attention distracted by a call on his cell phone. Murata Boy, the little white robotic cyclist, arrives on the scene to encourage the youngster. As the light from his electronic eyes reflects off the boy's face in a moment of intimate communion, the voiceover states simply, "Murata Boy is there for you." The pair ride off down the street, the father forgotten.

The ad clearly reflects a theme of all the dreams, from those of young manga readers to adult researchers—that humanoid robots are created for the purpose of helping people. The National Institute of Advanced Industrial Science and Technology, pursuing the publicly funded creation of a robot that will aid humans in everyday environments, expects that by 2025 robots will be helping seniors live independently by carrying out household chores. They've already spent seven years developing a prototype named HRP-2 Promet.

Samurai are not extinct in Japan. Kiyomori (opposite) is the flagship robot of Tmsuk, a Kitakyushu robotics firm founded in 2000. Developed with Waseda University's Atsuo Takanishi Laboratory, Kiyomori has thirty-nine degrees of freedom and can fully extend its legs at the knees when walking, a close approximation of human locomotion. Its *kachu* armor was made by Marutake Sangyo Co., which was armorer for Akira Kurosawa's 1985 epic *Ran*.

The National Institute of Advanced Industrial Science and Technology is developing the five-foot HRP-2 Promet humanoid (right) as a platform to assist people. Though its exterior appearance was conceived by Gundam fantasy robot designer Yutaka Izubuchi, HRP-2 can perform down-to-earth tasks like helping install wall paneling and fetching drinks

Promet can respond to verbal summons, tuck a chair under a table, turn the TV on and even fetch a canned drink from the fridge. But was it made to look like an old man, in an effort to blend in at a seniors' home? No, if anything, it looks—head to toe—like a sleek and powerful Gundam warrior, once again proof that juvenile dreams die hard. Maybe those who receive its care in the future will even prefer it that way.

For traditional dreamers, there is the strikingly exotic Kiyomori, a six-foot tall robot that walks naturally—or at least as naturally as any human walks when dressed in full Heian-period samurai armor. When a news item features Kiyomori praying at a shrine dedicated to traffic safety, it is a bizarre sight, and yet one that seems oddly natural in Japan. Maybe roboticist Mori is right and robots do have souls.

But perhaps the biggest dreamer of all is Mitsuo Kawato, a Kyoto-based researcher at the head of the state-backed ATR Computational Neuroscience Laboratories who has already built a humanoid that learns by observation. It can dance, play tennis and beat humans at air hockey. He's also proposing a grand undertaking, which would take at least ten years and ¥500 billion: to build a humanoid robot with the sensory, physical and mental abilities of a five-year-old child. Kawato compares his "Atom Project" to the United States' Apollo Program, and sees similar spinoff technologies and advances. At the time of writing, while the Japanese government struggles under a mountain of public debt, it is unclear whether he will win the funds needed for his plan.

When I was a boy, I drew robots from my imagination. I remember a giant warrior robot with a striped headdress inspired by the ancient pharaohs, and I called it Egyptorux. I spent hours and hours happily illustrating make-believe machines. Eventually, though, I forgot about them, put my markers away and stopped sketching. But Japan's robot creators have never lost touch with their childhood visions. And as technology evolves at runaway speed, maybe that's something to cling to.

"When we consider where human-humanoid relations will take us," wrote the organizers of a robot retrospective in Japan, "the answer is

Murata Boy, Murata Manufacturing Co.'s twenty-inch robot cyclist, is based on factory automation technology. It uses gyro sensors and a chest-mounted flywheel to maintain its balance, even while stationary, and ultrasonic sensors to avoid obstacles. It can travel forward and backward, in circles and even on a rail. Heavily promoted in an ad campaign, Murata Boy is presented as a friend to kids learning to ride. Says one development team member: "This reminds me of the twenty-first century we dreamed of as children."

These radio-controlled Neon robots were created in 2003 by Tomotaka Takahashi of Kyoto's Robo Garage to mark Mighty Atom's birth, but also as an attempt to turn science fiction into fact—specifically to "narrow the gap between imaginary robots appearing in manga and real robots." Not only do they look the part with their rounded, cartoonish carbon shells, but electromagnets in their feet allow them to stride majestically like heroic robots of fantasy.

© Tezuka Productions

that we come from Atom and we are heading toward Atom."

The robot journey is not without its twists. Science fiction author Shinichi Hoshi often wrote about robots, and in the short story "The Capricious Robot" (1966), the wealthy protagonist, Mr. N., buys a robot from a scientist after hearing that it can do anything. On his island retreat for a month's vacation, the machine is the perfect servant, cooking, cleaning and even entertaining. But one day, it stops working, forcing Mr. N. to prepare his own meals. Later, it runs off, so Mr. N. digs a pit to snare it. Once recaptured, the robot starts to work again, only to eventually go berserk, chasing its master up a tree.

In the end, Mr. N. demands his money back from the scientist. The latter answers calmly, "Of course I can make a robot that doesn't break down or go berserk. But if you lived with such a robot for a month you'd grow fat and your mind would go soft. That wouldn't do you any good. So this sort of robot companion is far better for a man."

Hoshi's stories are full of such delicious little ironies and subtle commentary on the vanity of human wishes. They also illustrate how robots embody the unexpected. As often as they disappoint our naïve expectations of finding the perfect tool, servant or conversation partner, robots astound us. Even if we are familiar with the principles, programming and mechanisms by which Asimo can walk and talk, we are still dumbstruck when he steps up to shake our hand.

So why are robots so loved in Japan? Simply because they are simultaneously science and fiction. They are the bright yellow Wakamaru robots, like the one that told my fortune in a Marunouchi furniture store, or the others that wait patiently in homes around Japan for their owners to return. And they are the plastic Gundam warriors holding court in my local barbershop—fuel for distant flights of imagination.

It is hard not to believe that when artificial intellilgence in humanoid form becomes commonplace, it will in large part be due to Japan's roboticists, and that the Japanese public will be the first to open their homes and lives to new, and always surprising, friends.

SELECTED SOURCES

Aramata, Hiroshi. *Daitoa kagaku kitan*. Tokyo: Chikuma Shobo, 1991.

Asada, Minoru, et al. "Robot Ecosystems." *InterCommunication* No. 28, Spring 1999, pp. 65–78.

Asimov, Isaac. *I, Robot*. New York: Bantam Dell, 2004.

Bandai Co. *Bandai Factbook 2005*. Tokyo: Bandai, 2005.

Brooks, Rodney A. *Flesh and Machines: How Robots Will Change Us*. New York: Vintage Books, 2003.

Čapek, Karel. *R.U.R. (Rossum's Universal Robots)*. London: Penguin, 2004.

Dick, Philip K. *Blade Runner (Do Androids Dream of Electric Sheep)*. New York: Del Ray, 1982.

Fukuda, Toshio. *Tetsuwan atomu no robotto gaku*. Tokyo: Shueisha, 2003.

Hirai, Kazuo, et al. "The Development of the Honda Humanoid Robot." *Proceedings of the IEEE International Conference on Robotics & Automation*, 1998, pp. 1321–1326.

IDG Japan. *Robo Next: Saishin robotto katarogu 2005*. Tokyo: IDG Japan, 2005.

Inoue, Haruki. *Nihon robotto soseiki 1920–1938*. Tokyo: NTT Shuppan, 1993.

Ishiguro, Hiroshi, et al. "Evaluating Humanlikeness by Comparing Responses Elicited by an Android and a Person." *Proceedings of the Second International Workshop on Man-Machine Symbiotic Systems*, November 2004, pp. 373–383.
—"Generating Natural Motion in an Android by Mapping Human Motion." *Proceedings of IEEE/RSJ International Conference on Intelligent Robots and Systems*, August 2005, pp. 1089–1096.

Kato, Ichiro. *Development of Waseda Robot: The Study of Biomechanisms at Kato Laboratory*. Tokyo: Waseda Humanoid Robotics Institute, 1985.

Kretschmer, Angelika. "Mortuary Rites for Inanimate Objects: The Case of Hari Kuyou." *Japanese Journal of Religious Studies*, Fall 2000, pp. 380–404.

Kurzweil, Ray. *The Age of Spiritual Machines: When Computers Exceed Human Intelligence*. New York: Penguin, 1999.

Moravec, Hans. *Robot: Mere Machine to Transcendent Mind*. New York: Oxford University Press, 1999.

Mori, Masahiro. *The Buddha in the Robot: A Robot Engineer's Thoughts on Science and Religion*. Trans. Charles S. Terry. Tokyo: Kosei, 1981.

Murakami, Takashi, ed. *Little Boy: The Arts of Japan's Exploding Subculture*. New Haven: Yale University Press, 2005.

Nagai, Go. *Majinga zetto orijinaru ban*. Tokyo: Kodansha, 1999.

National Museum of Modern Art, Tokyo, ed. *Osamu Tezuka Exhibition*. Tokyo: National Museum of Modern Art, Tokyo, 1990.

New Energy and Industrial Technology Development Organization (NEDO). *Expo 2005 Aichi Robot Project Guidebook*. Kawasaki: NEDO, 2005.

Okubo, Hironori, et al. *The Robot Chronicles: Tetsuwan atomu no kisekiten*. Tokyo: Asahi Shimbun, 2002.

Patten, Fred. *Watching Anime, Reading Manga: 25 Years of Essays and Reviews*. Berkeley: Stone Bridge Press, 2004.

Sadamoto, Yoshiyuki. *Neon Genesis Evangelion*. Trans. Mari Morimoto. San Francisco: Viz Communications, 2004.

Sakai, Tadayasu, et al. *The Čapek Brothers and Czech Avant-gardes*. Tokyo: IDF, 2002.

Schodt, Frederik L. *Manga! Manga! The World of Japanese Comics*. Tokyo: Kondansha International, 1983.
—*Inside the Robot Kingdom: Japan, Mechatronics, and the Coming Robotopia*. Tokyo: Kodansha International, 1988.
—*Dreamland Japan: Writings on Modern Manga*. Berkeley: Stone Bridge Press, 1996.

Schreiber, Mark. "Karakuri Ningyo: The Amazing Ancestors of Today's Industrial Robots," *Japan Close-Up*, pp. 14–20, November 2005.

Sena, Hideaki. *Robotto opera*. Tokyo: Kobunsha, 2004.
—"Astro Boy was Born on April 7, 2003." *Japan Echo*, Vol. 30, No. 3, pp. 9–12, August 2003.

Tezuka, Osamu. *Tetsuwan atomu 1*. Tokyo: Kodansha, 2002.
—*Astro Boy vol. 15*. Trans. Frederik L. Schodt, Milwaukie, OR: Dark Horse Manga, 2003.

Ueda, Kenji. *Shinto*. Tokyo: Jinja Honcho, 1999.

Yokoyama, Mitsuteru. *Tetsujin nijuhachigo: Gensaku kanzenban*. Tokyo: Hikari Productions, 2005.

ONLINE RESOURCES

* indicates websites only in Japanese

Wakamaru www.mhi.co.jp/kobe/wakamaru/english
Gakken karakuri dolls
   shop.gakken.co.jp/otonanokagaku/karakuri.html *
Hisashige Tanaka www.toshiba.co.jp/spirit/en/index.html
Tezuka Osamu World www.tezuka.co.jp
Bandai www.bandai.co.jp
Sunrise www.sunrise-inc.co.jp
Gainax www.gainax.co.jp *
Sakakibara Kikai www.sakakibara-kikai.co.jp *
Waseda University Humanoid Robotics Institute
   www.humanoid.waseda.ac.jp
Aibo www.sony.net/Products/aibo
Qrio www.sony.net/SonyInfo/QRIO/top_nf.html
Paro paro.jp/english/index.html
Business Design Laboratory www.business-design.co.jp

Kondo Kagaku www.kopropo.co.jp *
ZMP www.zmp.co.jp
Asimo world.honda.com/ASIMO
Fujitsu Automation www.automation.fujitsu.com
Toyota Partner Robots www.toyota.co.jp/en/special/robot
RoboCup www.robocup.org
Vision Nexta www.vstone.co.jp/top/products/robot/v2/ *
Robo Garage www.robo-garage.com
Osaka University Intelligent Robotics Laboratory
   www.ed.ams.eng.osaka-u.ac.jp
Kokoro www.kokoro-dreams.co.jp
Murata Manufacturing www.murata.com
Tmsuk www.tmsuk.co.jp
AIST www.aist.go.jp

ACKNOWLEDGMENTS

This work is dedicated to the memory of my father, who built things by the sweat of his brow.

It is also in a sense a tribute to the vision and perseverance of the many robot creators in Japan, from manga artists to mechanical engineers, who are pursuing the robot dream in their own unique way. I salute their boundless imagination, fearless attitude and innovative derring-do. I have no doubt that their robotic creations will continue to inspire people everywhere.

I would like to express special gratitude to my editor Gregory Starr for his faith in the project, Yoshino Matsui for her invaluable practical and moral support, and my family.

My sincere thanks also go to the following for their kind assistance: Art Director Kazuhiko Miki, Tony McNicol, Yuji Koga, Masahiko Kitamura, Hiroe Teramura, Tomotaka Takahashi, Hiroshi Ishiguro, Karl MacDorman, Yoshikazu Suematsu, Jozef Kelemen, Masahito Iwano, Yutaka Izawa, Hiroshi Araki, Haruyuki Nakano, Mark Schreiber, Gregory McCartney, Tony Jebara and Kevin Malakoff. I would also like to thank the many companies, organizations and individuals that cooperated in the production of this book. And to the humanoid machines of Japan, real and unreal, past, present and future, I say: *Domo arigato, Mr. Robotto*.

Timothy N. Hornyak
Tokyo, 2006

PHOTO CREDITS

PAGE 6, 8, 10, 11 Mitsubishi Heavy Industries Ltd., *14, 15* (4) Timothy N. Hornyak; *18* Mizuho Kuwata; *20, 21* Hornyak; *22, 23, 24* Mizuho Kuwata; *25* Kurume City Board of Education; *26* (TOP) Toyota Collection, (BOTTOM) Japan Karakuri-automata Society: Susumu Higashino; *27* (TOP) Toyota Collection; (BOTTOM) National Science Museum; *30–31* Hornyak; *32* Tsubouchi Memorial Theatre Museum, Waseda University F32-064-1; *33* Hayuza Company/ Poster design by Jun'ichi Takei; *34, 35* Hornyak; *36, 37* (3) from *Daitoa kagaku kitan* by Hiroshi Aramata, published by Chikuma Shobo, 1991; *38* Masahito Iwano; *39 : 50–51* Mizuho Kuwata (art by Hiroshi Araki, from the collection of Yutaka Izawa); *54–55* (5) Mizuho Kuwata (from the collection of Yutaka Izawa); *58* © Hikari Production/Shikishima Juko; *59* © Hikari Production; *60, 61* © Dynamic Productions; *62, 64, 65* Sunrise Inc.; *66–67* © 2005 Tenmyouya Hisashi/Sotsu Agency/Sunrise Inc., photograph by Keizo Kioku; *68* Hornyak; *69* © Sotsu Agency/Sunrise/MBS; *70, 71* © Gainax Inc.; *74* (2), *75, 76* (2), *77* (3), *78* courtesy Waseda University Humanoid Robotics Institute; *79, 81, 83, 84* Hornyak; *86, 87* Sony; *88* © National Institute of Advanced Industrial Science and Technology (AIST); *89* Hornyak; *90* Business Design Laboratory Co., Ltd.: *91* Business Design Laboratory Co., Ltd./SANRIO; *92, 93* (3) Kondo Kagaku Co. Ltd.; *94* (3), *95, 98, 99* © 2006 ZMP Inc.; *96* ZMP; *97* Asahi Shimbun; *102* (7), *103* (7), *104, 105, 106* Honda Motor Co. Ltd.; *107* Lidové Noviny; *108, 109* (2) Honda Motor Co. Ltd ; *110, 111, 112, 113* Sony; *114* © Fujitsu Frontech Limited; *115* (2) JVC; *116* Toyota Motor Corporation; *118, 119* Vstone Co., Ltd.; *120* © The RoboCup Federation, *121, 122, 123* (2) Hornyak; *124* © The RoboCup Federation; *125* Hornyak; *126* (3) Tomotaka Takahashi (1) Hornyak; *127* (3) Tomotaka Takahashi; *128, 129* © The RoboCup Federation; *130–131* (3) Hornyak; *134* Hiroshi Ishiguro Laboratory; *135, 136, 139* Shuichi Yamagata; *140, 142, 143, 145, 146* Hornyak; *150–151* Kitano Symbiotic Systems Project, Japan Science and Technology Corporation; *152* tmsuk Co. Ltd., *153* AIST/Kawada Industries, Inc.; *154–155* (2) Murata Manufacturing Co., Ltd.; *156* Tomotaka Takahashi/Robo Garage

（英文版）ロボット
Loving the Machine

2006 年5月25日　第 1 刷発行

著　者　　ティモシー・N・ホーニャック
発行者　　富田 充
発行所　　講談社インターナショナル株式会社
　　　　　〒112-8652 東京都文京区音羽 1-17-14
　　　　　電話　03-3944-6493（編集部）
　　　　　　　　03-3944-6492（マーケティング部・業務部）
　　　　　ホームページ　www.kodansha-intl.com

印刷・製本所　大日本印刷株式会社

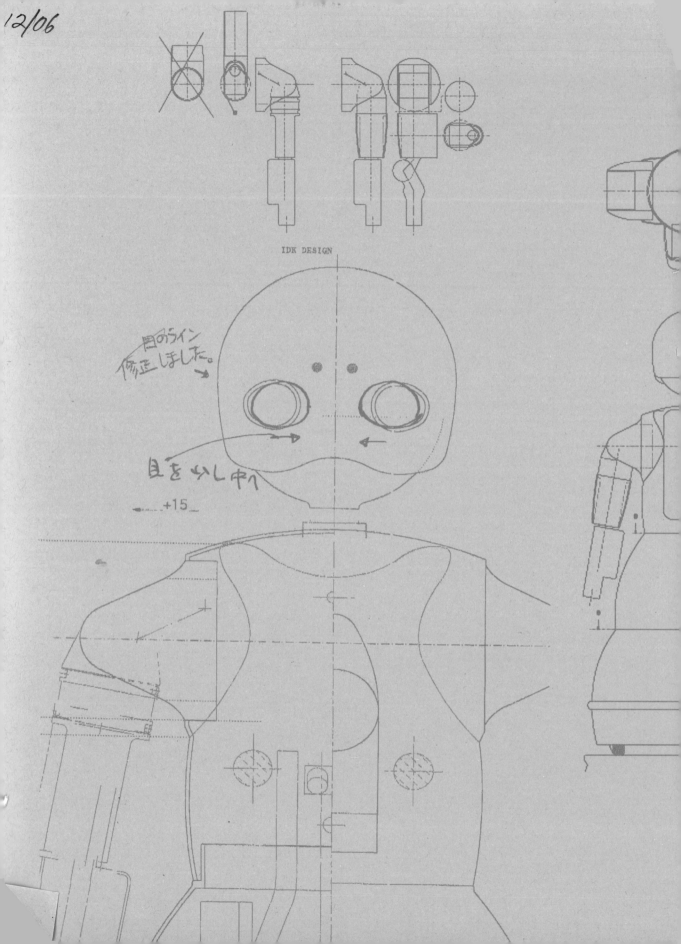